# Protein folding *in silico*

Published by Woodhead Publishing Limited, 2012

# Woodhead Publishing Series in Biomedicine

1 Practical leadership for biopharmaceutical executives
*J. Y. Chin*
2 Outsourcing biopharma R&D to India
*P. R. Chowdhury*
3 Matlab® in bioscience and biotechnology
*L. Burstein*
4 Allergens and respiratory pollutants
*Edited by M. A. Williams*
5 Concepts and techniques in genomics and proteomics
*N. Saraswathy and P. Ramalingam*
6 An introduction to pharmaceutical sciences
*J. Roy*
7 Patently innovative: How pharmaceutical firms use emerging patent law to extend monopolies on blockbuster drugs
*R. A. Bouchard*
8 Therapeutic protein drug products: Practical approaches to formulation in the laboratory, manufacturing and the clinic
*Edited by B. K. Meyer*
9 A biotech manager's handbook: A practical guide
*Edited by M. O'Neill and M. H. Hopkins*
10 Clinical research in Asia: Opportunities and challenges
*U. Sahoo*
11 Therapeutic antibody engineering: Current and future advances driving the strongest growth area in the pharma industry
*W. R. Strohl and L. M. Strohl*

Published by Woodhead Publishing Limited, 2012

27 Clinical research in paediatric psychopharmacology: A practical guide
*P. Auby*

28 The application of SPC in the pharmaceutical and biotechnology industries
*T. Cochrane*

29 Ultrafiltration for bioprocessing
*H. Lutz*

30 Therapeutic risk management of medicines
*A. K. Banerjee and S. Mayall*

31 21st century quality management and good management practices: Value added compliance for the pharmaceutical and biotechnology industry
*S. Williams*

32

33 CAPA in the pharmaceutical and biotech industries
*J. Rodriguez*

34 Process validation for the production of biopharmaceuticals: Principles and best practice.
*A. R. Newcombe and P. Thillaivinayagalingam*

35 Clinical trial management: An overview
*U. Sahoo and D. Sawant*

36 Impact of regulation on drug development
*H. Guenter Hennings*

37 Lean biomanufacturing
*N. J. Smart*

38 Marine enzymes for biocatalysis
*Edited by A. Trincone*

39 Ocular transporters and receptors in the eye: Their role in drug delivery
*A. K. Mitra*

40 Stem cell bioprocessing: For cellular therapy, diagnostics and drug development
*T. G. Fernandes, M. M. Diogo and J. M. S. Cabral*

41

42 Fed-batch fermentation: A practical guide to scalable recombinant protein production in *Escherichia coli*
*G. G. Moulton and T. Vedvick*

43 The funding of biopharmaceutical research and development
*D. R. Williams*

44 Formulation tools for pharmaceutical development
*Edited by J. E. A. Diaz*

45 Drug-biomembrane interaction studies: The application of calorimetric techniques
*R. Pignatello*

46 Orphan drugs: Understanding the rare drugs market
*E. Hernberg-Ståhl*

47 Nanoparticle-based approaches to targeting drugs for severe diseases
*J. L. A. Mediano*

48 Successful biopharmaceutical operations
*C. Driscoll*

49 Electroporation-based therapies for cancer
*Edited by R. Sundarajan*

50 Transporters in drug discovery and development
*Y. Lai*

51 The life-cycle of pharmaceuticals in the environment
*R. Braund and B. Peake*

52 Computer-aided applications in pharmaceutical technology
*Edited by J. Petrović*

53 From plant genomics to plant biotechnology
*Edited by P. Poltronieri, N. Burbulis and C. Fogher*

54 Bioprocess engineering: An introductory engineering and life science approach
*K. G. Clarke*

55 Quality assurance problem solving and training strategies for success in the pharmaceutical and life science industries
*G. Welty*

56 Nanomedicine: Prognostic and curative approaches to cancer
*K. Scarberry*

57 Gene therapy: Potential applications of nanotechnology
*S. Nimesh*

58 Controlled drug delivery: The role of self-assembling multi-task excipients
*M. Mateescu*

Woodhead Publishing Series in Biomedicine: Number 22

# Protein folding *in silico*

## Protein folding versus protein structure prediction

Edited by
Irena Roterman-Konieczna

WP
WOODHEAD
PUBLISHING

Oxford Cambridge Philadelphia New Delhi

Published by Woodhead Publishing Limited, 2012

Woodhead Publishing Limited, 80 High Street, Sawston, Cambridge, CB22 3HJ, UK
www.woodheadpublishing.com
www.woodheadpublishingonline.com

Woodhead Publishing, 1518 Walnut Street, Suite 1100, Philadelphia, PA 19102-3406, USA

Woodhead Publishing India Private Limited, G-2, Vardaan House, 7/28 Ansari Road, Daryaganj, New Delhi – 110002, India
www.woodheadpublishingindia.com

First published in 2012 by Woodhead Publishing Limited
ISBN: 978-1-907568-17-6 (print); ISBN 978-1-908818-25-6 (online)
Woodhead Publishing Series in Biomedicine ISSN 2050-0289 (print);
ISSN 2050-2097 (online)



British Library Cataloguing-in-Publication Data: A catalogue record for this book is available from the British Library.

Library of Congress Control Number: 2012942498



Typeset by RefineCatch Limited, Bungay, Suffolk

To my best teachers:

Kindergarten – Katarzyna Wędzicha

Elementary school – Jadwiga Baluch

Secondary school – Wojciech Komusiński

Jagiellonian University – Professor Kazimierz Gumiński

Published by Woodhead Publishing Limited, 2012

# Contents





Published by Woodhead Publishing Limited, 2012

Published by Woodhead Publishing Limited, 2012

Published by Woodhead Publishing Limited, 2012

Published by Woodhead Publishing Limited, 2012

# List of figures





Published by Woodhead Publishing Limited, 2012

Published by Woodhead Publishing Limited, 2012

Published by Woodhead Publishing Limited, 2012

Published by Woodhead Publishing Limited, 2012

# List of tables





Published by Woodhead Publishing Limited, 2012

Published by Woodhead Publishing Limited, 2012

# About the editor

Irena Roterman-Konieczna is the leader of the bioinformatics group at the Jagiellonian University, Kraków, Poland, and is Professor of Bioinformatics and the Head of the Department of Bioinformatics and Telemedicine in the Medical College at the Jagiellonian University. She graduated from the Jagiellonian University in 1974, majoring in theoretical chemistry. This provided a good basis on which to study advanced bioinformatics techniques during her academic career. Her PhD thesis (submitted and defended at the Medical Academy in Kraków) involved the simulation of the hinge region in recombinant hinge region immunoglobulin. A two-year collaboration with Harold A. Scheraga's group at Cornell University provided an excellent introduction to high-level bioinformatics oriented on protein structure prediction. During her stay at Cornell, she coauthored two papers with Harold A. Scheraga, presenting a detailed comparative analysis of the popular software packages Amber, Charmm and ECEPP.

Experience gained at Cornell University enabled her to establish her own bioinformatics research team following her return to Kraków. The group focused on amyloid analysis, particularly in the context of immunoglobulins and serpins, as well as on the application of amyloid markers, such as Congo red. This work resulted in a hypothesis regarding the supramolecular structure of Congo red. Further collaboration with the experimental immunochemistry



Published by Woodhead Publishing Limited, 2012

group at the Jagiellonian University Medical College Institute of Medical Biochemistry resulted in the identification of structural motifs responsible for immunological signal transduction in immunoglobulins.

Irena Roterman-Konieczna's research into protein structure prediction yielded a new model for simulating the protein folding process, which attempts to account for observational phenomena when one is constructing *in silico* simulations. Progess towards this goal is advanced by introducing intermediate folding steps. The early-stage structural form of the polypeptide chain is generated solely on the basis of analysis of peptide bond planes. A geometric model has been developed where all polypeptide chains are treated as helical structures with different curvature radii (the $\beta$-structural form is assumed to represent a helix with a very large radius of curvature). This interpretation enables structural changes to be treated as a continuous process. Another consequence of the model is the establishment of a limited conformational subspace which, in mechanistic terms, defines the conformation of the early-stage intermediate. Thus, the model presented solves the problem of defining a starting structure for energy optimization algorithms.

In contrast to the first step, the second stage of the protein folding process depends on the influence of the external environment, namely water. The "oil drop" model, originally introduced by W. Kauzmann, has been extended into a "fuzzy oil drop" model by applying a 3D Gaussian function to describe the structure and behavior of polypeptide chains with variable hydrophobicity. Polypeptide chain folding, where hydrophobic residues are directed towards the center of the globule, with simultaneous exposure of hydrophilic residues on its surface, follows the hydrophobicity distribution expressed by the 3D Gaussian function. This function predicts the emergence of a perfect hydrophobic



Published by Woodhead Publishing Limited, 2012

core, thus producing a highly soluble protein molecule. Identification of proteins whose structure remains in good agreement with the "fuzzy oil drop" model (e.g., antifreeze proteins and downhill proteins) supports the model's validity. The observed hydrophobicity irregularities (deviations from the idealized model, as predicted by the Gaussian function) seem to be related to the protein's biological function (ligand binding, protein complexation, etc.).

Irena Roterman-Konieczna's involvement in medical bioinformatics covers the analysis of aspirin metabolites, conducted in collaboration with medical practitioners at Jagiellonian University Medical College clinics. She also collaborates with the Faculty of Physics, Astronomy and Applied Computer Science, organizing bioinformatics courses for students of applied computer science and biophysics. She has supervised numerous master's and doctoral theses. Her PhD students are pursuing successful careers at American and European universities, participating in many research projects.

Irena Roterman-Konieczna is also an experienced participant in EU-funded research activities, including collaborations with China.

The CYFRONET AGH Academic Computer Center is one of the foremost Polish supercomputing centres, currently ranked 60th in the worldwide TOP-500 computing-power list. Irena Roterman-Konieczna's collaboration with CYFRONET dates back to the establishment of the Center. Tools implemented in accordance with Irena's models are being deployed in the national grid computing infrastructure (PL-Grid), which aims to provide computational and software resources to Polish researchers.

Irena Roterman-Konieczna is the editor-in-chief of the scientific journal *Bio-Algorithms and Med-Systems*, which has been invited to join the deGruyter journal family, starting



Published by Woodhead Publishing Limited, 2012

in January 2012. The journal (which was originally established by Irena) aims to forge a link between the medical community and cutting-edge computational technologies in the context of molecular and macro-level models for theoretical research and medical practice. Its website is *http://www.bams.cm-uj.krakow.pl.*

Professor Irena Roterman-Konieczna is the current Dean of the Department of Bioinformatics and Telemedicine, Jagiellonian University Medical College, Kraków, Poland.

*Contact details*

Irena Roterman-Konieczna
Department of Bioinformatics and Telemedicine
Jagiellonian University Medical College
Lazarza 16, 31-530 Kraków, Poland
Email: *irena.roterman-konieczna@uj.edu.pl*



Published by Woodhead Publishing Limited, 2012

# List of contributors

Paweł Alejster, Master's student, Jagiellonian University, Faculty of Physics, Astronomy and Applied Computer Science, Kraków, Poland

Mateusz Banach, PhD student, Jagiellonian University, Faculty of Physics, Astronomy and Applied Computer Science, Kraków, Poland

Zbigniew Baster, Master's student, AGH – University of Science and Technology, Faculty of Physics and Applied Computer Science, Kraków, Poland

Dawid Dułak, Master's student, AGH – University of Science and Technology, Faculty of Physics and Applied Computer Science, Kraków, Poland

Wiktor Jurkowski, Doctor of Chemistry, Luxembourg Centre for Systems Biomedicine, University of Luxembourg

Leszek Konieczny, Emeritus Professor of Biochemistry, Chair of Medical Biochemistry, Medical College, Jagiellonian University, Kraków, Poland

Marcin Król, Doctor of Chemistry, Department of Bioinformatics and Telemedicine, Medical College, Jagiellonian University, Kraków, Poland



Published by Woodhead Publishing Limited, 2012

Damian Marchewka, PhD student, Jagiellonian University, Faculty of Physics, Astronomy and Applied Computer Science, Kraków, Poland

Monika Piwowar, Doctor of Genomics, Department of Bioinformatics and Telemedicine, Medical College, Jagiellonian University, Kraków, Poland

Katarzyna Stapor, Professor of Computer Science, Silesian University of Technology, Gliwice, Poland

Zdzisław Wiśniowski, Master of Computer Science, Department of Bioinformatics and Telemedicine, Medical College, Jagiellonian University, Kraków, Poland

Wojciech Ziajka, Master of Computer Science, Department of Bioinformatics and Telemedicine, Medical College, Jagiellonian University, Kraków, Poland



Published by Woodhead Publishing Limited, 2012

# Introduction

Much has been said and written about the progress that has been made in the natural sciences over the past several decades. The breakthroughs that have been made in biology carry obvious scientific benefits, while at the same time contributing to practical applications in the medical, veterinary, and environmental sciences and many other areas.

The history of experimental science goes back thousands of years, from simple observations of naturally occurring structures and phenomena to the analysis of living organisms on a molecular level with the aid of specialized laboratory equipment of high technological quality. Quite recently (around the 1980s), an entirely new scientific discipline has emerged: the application of computational technologies to the study of natural processes. This research field is frequently referred to as "bioinformatics."

The ability to sequence and catalog the genetic code of various organisms, from simple bacteria and viruses all the way to mammals, including *Homo sapiens* in particular, is an invaluable tool in the hands of modern biologists. Yet every experiment has its limits – whether because of technical limitations (e.g., the resolution of imaging techniques) or as a result of the inapplicability of useful research methods (such as crystallization) to certain biological structures (e.g., immunoglobin, which consists of loosely connected subunits of comparable size and does not possess the conformational



Published by Woodhead Publishing Limited, 2012

uniformity required to form a crystal). In some cases, the high speed of biological processes or the complex environment in which they occur renders experimentation impossible or impractical. In order to transcend such limits, biology has enlisted the aid of computational science. The role of computers in biology was initially subordinate and limited to managing large volumes of data (such as those generated by the Human Genome Project); however, at some point, the computer operators and soon-to-be bioinformatics experts began to ask scientific questions of their own. The first such questions involved "reading and interpreting" the content of nucleotide sequences by identifying genes and comparing various fragments of genetic code. As time has gone by, bioinformatics has evolved from a purely technical occupation to a science in its own right, developing new scientific problems to be solved rather than simply verifying existing hypotheses.

It is expected that numerical models must therefore go beyond capturing physical phenomena – indeed, they should provide answers to questions such as "why does this particular phenomenon occur?" and "why does it occur the way it does?" Bioinformatics is tasked with finding theoretical solutions to problems that cannot be approached by experimentation alone. Moreover, it should be able to devise new experiments that will shed light on not entirely understood mechanisms and highlight new problems that would not otherwise attract attention.

One of the major challenges faced by bioinformatics is to provide a theoretical foundation for protein structure prediction. The structure of a folded protein depends on its underlying sequence of amino acids and, at the same time, determines the protein's role in biological systems (which is another important area of study). If a cell is capable of consistently generating proteins with suitable structures, encoded in the genetic material in the nucleus, it is natural to expect scientists to be able to simulate this process *in silico*.

This somewhat generic problem can be divided into two distinct challenges. The first is to predict the structure of a folded protein on the basis of what might be called an "educated guess." The second is to gain insight into the folding process itself and then apply it to obtain the correct three-dimensional structure for a known sequence of amino acids. The latter approach is sometimes called the "*ab initio*" or "new fold" approach.

Clearly, the most welcome result of scientific study of this problem would be to explain why proteins fold the way they do (as opposed to methods limited solely to predicting their structure). Such knowledge would enable us not only to simulate folding but also to guide it in desirable ways, which would precipitate breakthroughs in drug design (such as the popular notion of "designer enzymes"). At the same time, theoretical knowledge of the folding process would explain many common pathologies and suggest ways to ameliorate them. The goal of this book is therefore to propose and evaluate a theoretical model that enables simulation of the protein folding process. Our model attempts to incorporate, as far as possible, the results of experimental studies and exploit knowledge about the mechanisms that affect naturally occurring proteins. The intention is to simulate each stage in the folding process rather than simply predict its final result. Readers are invited to judge to what extent this objective has been successfully accomplished.

This brief introduction to the protein folding problem would not be complete without a reference to CASP (Critical Assessment of protein Structure Prediction, *http://predictioncenter.org/*), a project aimed at monitoring progress in computerized protein structure prediction tools. Ever



Published by Woodhead Publishing Limited, 2012

since 1994, CASP has been diligent in describing the state of the art and tracking scientific progress in this field.

Adapting our model to experimental data requires us to first divide the folding process into several stages, reflecting the way it occurs in nature [1]. Obviously, the simplest possible approach to protein structure prediction would be to express it as follows:

$$U(AA) \rightarrow N$$

where U is the unstructured form, N is the native structure, and AA is the amino acid sequence. This simplistic "guesswork" model does not concern itself with anything that happens during folding – its only goal is to produce the correct result. Practical implementations of such an approach include simulations of *N*-atom systems in search of a global energy minimum. Indeed, if efficient algorithms for minimizing complex functions were available (which, unfortunately, is not the case), the entire folding process could conceivably be expressed by this simple formula and the sole task of *in silico* tools would be to find a global minimum of the free-energy function representing the molecule under consideration.

The guesswork model, while simple, is misleading. Experimental data suggests that the folding process progresses through several distinct intermediate stages (the number of which is undetermined and may depend on the specific properties of a given protein):

$$U \rightarrow I_1 \rightarrow I_2 \rightarrow \ldots I_n \rightarrow N$$

where $I_n$ is an intermediate form with cardinality $n$ (where $n$ is not known). Our model proposes exactly two intermediates:

$$U \rightarrow ES \rightarrow LS \rightarrow N$$

where ES is the early-stage intermediate, and LS is the late-stage intermediate.



Published by Woodhead Publishing Limited, 2012

The following chapters will discuss the basic assumptions applicable to each intermediate form and explain how they affect the process of generating protein complexation sites, ligand-binding pockets, enzymatic active sites, etc. The conclusions (based on analysis of protein structures in the context of the proposed model) suggest that it might be useful (and necessary) to introduce additional intermediate forms, which brings us closer to our goal – i.e., answering the question "why do proteins fold the way they do?"

*Irena Roterman-Konieczna*
*Jagiellonian University Medical College Kraków,*
4 January 2012

## Reference

1. Creighton T.E. (1990) Protein folding. *Biochem. J.* 270, 1–16.



Published by Woodhead Publishing Limited, 2012

1

# The early-stage intermediate

*W. Jurkowski, Z. Baster, D. Dułak, and I. Roterman-Konieczna*

**Abstract:** The multistep polypeptide chain folding model presented in this chapter involves several intermediates, the first of which is called the early-stage (ES) intermediate. This intermediate is assumed to be defined solely on the basis of the backbone conformation and does not take side chains into account. The geometric principles that guide the backbone alignment process and its quantitative influence on the structural arrangement of the folded chain can be expressed by means of a contingency table, linking known structural motifs to specific polypeptide sequences. The basic unit of this algorithm is the tetrapeptide, and the corresponding ES conformational subspace is assumed to consist of seven types of motifs. This limited subspace represents a subset of the full conformational space (i.e., the Ramachandran plot). The volumetric structure of the ES intermediate corresponds to the output of the early folding stage and, simultaneously, provides input for further stages of the folding process.

**Key words:** early-stage intermediate, backbone structure, limited conformational subspace.



Published by Woodhead Publishing Limited, 2012

## 1.1 Geometric model

The existence of a limited conformational subspace for early-stage folding intermediates was postulated by Alonso and Daggett.[1] The model presented in this book seems to satisfy this expectation.

The basic assumption underpinning the early-stage (ES) model is that the initial steps in the process of polypeptide chain folding are almost entirely determined by the backbone, which can be plotted as a sequence of flat rhomboids, each representing the conformation of a single peptide bond plane. The resulting chain can subsequently be treated as a series of spatially aligned "tiles" (or dipoles). One of the standard primary structural motifs is the $\alpha$-helix, in which consecutive tiles (i.e., peptide bonds) follow a common direction, with the angle between each pair of tiles approaching 0°. This is evidenced by the ideal matching of N–H and C=O moieties, resulting in a regular lattice of variable-distance hydrogen bonds. A different situation occurs in the case of the $\beta$-sheet structure, where individual tiles are arranged in an antiparallel fashion with the angle between each pair of tiles close to 180°. Such an alignment also permits a lattice of hydrogen bonds to form between neighboring chains.[2,3]

Given this model, we can define a parameter (called the V-angle) that reflects the angle between each pair of adjacent tiles. The V-angle approaches 0° in $\alpha$-helices, while remaining close to 180° in $\beta$-sheet structures. All other structures (which are more or less unstructured) adopt V-angles in the range of 0–180°.

V-angle variations are a simple consequence of changes in the dihedral angles in each peptide. Taken as a whole, these values unambiguously determine the structure of the entire protein. An important issue here is to identify the relation between the V-angle values and the structural properties of



Published by Woodhead Publishing Limited, 2012

the protein. The solution lies in visual assessment of the $\alpha$-helix and $\beta$-sheet structures. The V-angle determines the chain's curvature radius, which may range from small (helix) to large (sheet). An "ideal" $\beta$-structural fragment would be characterized by an infinitely large curvature, resulting in a perfectly linear structure, although in reality $\beta$-sheets exhibit a finite (though large) curvature.

A consequence of the model presented here is that each polypeptide chain may be formally described as a spring (helix) with a variable curvature. In order to quantitatively validate this approach, a structural analysis of a pentapeptide chain was performed, where the angles $\Phi$ and $\Psi$ were altered stepwise in 5° increments to cover the entire Ramachandran plot (i.e., the protein's conformational space). (The dihedral angle $\Phi$ expresses the state of free rotation of the N–$C_\alpha$ bond in the peptide bond, and the dihedral angle $\Psi$ the state of free rotation of the $C_\alpha$–C bond.) For each structure generated in this fashion, the V-angle $V$ and the radius of curvature $R$ were duly calculated.

Determining these two parameters requires the definition of a specific spatial alignment as a reference. This can be done by averaging the positions of the carbons and oxygens belonging to the carboxyl groups that determine the $Z$ axis. The averaged position of all C atoms coincides with the origin of the coordinate system, and the averaged position of the O atoms is localized on the $Z$ axis. The positions of the $C_\alpha$ atoms, when projected onto the $XY$ plane, determine the radius of curvature of the pentapeptide. For this orientation, the radius of curvature $R$ can be calculated on the basis of the $C_\alpha$ positions on the $XY$ plane.

Plotting each pair of $V$ and $R$ values reveals a parabolic dependency, as depicted in Figure 1.1(b). Note that for the sake of clarity, the $R$ scale is logarithmic (to accommodate the large radii of $\beta$-sheet structures). Figure 1.1(b) presents a



Published by Woodhead Publishing Limited, 2012

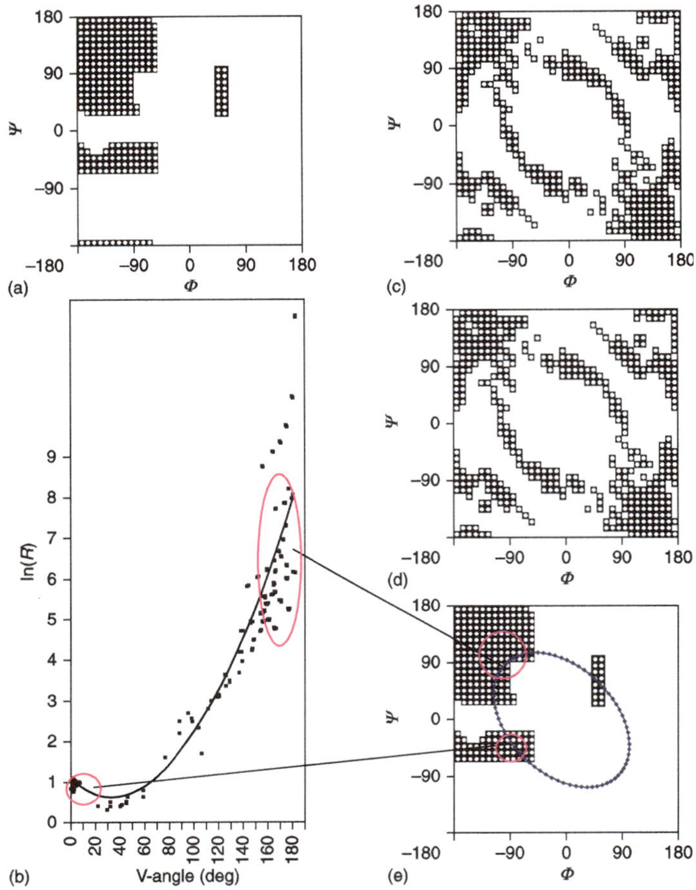


**Figure 1.1** Model of the early-stage intermediate. (a) Ramachandran plot, with low-energy structures indicated; (b) relation between ln(*R*) and *V* for structures belonging to the low-energy zone of the Ramachandran plot (as shown in (a)); (c) distribution of structures that exhibit the sought-after relation between ln(*R*) and *V*; (d) elliptical path representing the structures identified in (b); (e) relation between the elliptical path and low-energy areas corresponding to specific secondary structural motifs. The arrows indicate correspondences between points on the Ramachandran plot (d) and ln(*R*)/*V* values (b)



Published by Woodhead Publishing Limited, 2012

dependency chart for the low-energy areas of Figure 1.1(a). Its approximation function can be expressed as a second-degree polynomial,

$$\ln(R) = 0.00034\, V^2 - 0.02009\, V + 0.848 \qquad [1.1]$$

Assuming that the above equation accurately reflects the relation between the V-angle and the structural radius of curvature, the following question can be asked: which section of the Ramachandran plot do the early-stage intermediates belong to? As it turns out, those values which satisfy Eq. 1.1 (Figure 1.1(c)) point to specific areas where the V-angles and the corresponding radii remain in agreement with the approximation presented here. A cursory glance at the resulting plot reveals that the simplest way to connect all such areas (linking secondary protein structures) is with an elliptical path (Figure 1.1(d)).[4] The equation of the ellipse is as follows:

$$\begin{cases} \varphi = -A\cos(t) - B\sin(t) \\ \psi = A\cos(t) - B\sin(t) \end{cases} \qquad [1.2]$$

where $A$ and $B$ are the long and short diagonals, respectively.

The general conclusion is that the spectrum of ES folding intermediates (in which the conformation of the backbone plays a decisive role) can be described by an ellipse, depicting the optimal relation between the values of $V$ and $R$ for the polypeptide chain. From now on, this elliptical path will be taken to define the limited conformational subspace for the ES folding intermediate.

## 1.2 Structural alphabet

If we assume that the elliptical path presented in Figure 1.2 represents the full spectrum of ES structural variability in the



Published by Woodhead Publishing Limited, 2012

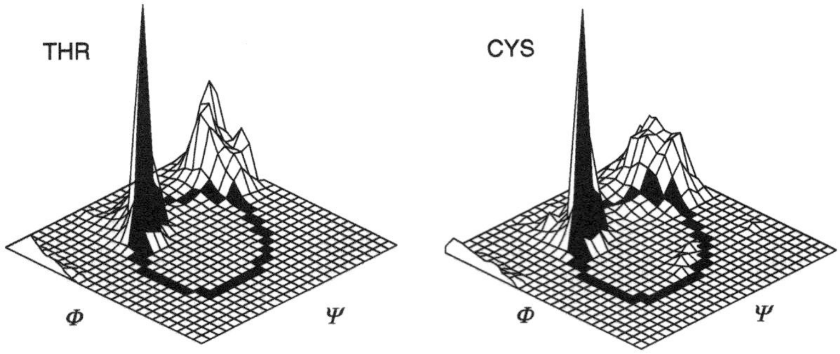


**Figure 1.2** **Elliptical path (black) drawn on a probability distribution plot representing the angles $\Phi$ and $\Psi$ for arbitrarily selected THR (left) and CYS (right) peptides. The sharp spike corresponds to a-helical structures, and the broader peak indicates $\beta$-conformations**

polypeptide chain, we can transform the native structures found in the Protein Data Bank (PDB) into their corresponding ES structural forms. This step requires the calculation of equivalent values $\Phi_e$ and $\Psi_e$ for each pair of angles $\Phi$ and $\Psi$ in the polypeptide chain by applying a least-distance rule between the protein's native form and its structure as it appears in the limited conformational subspace (represented by the elliptical path).

In mathematical terms, the operation presented here implies finding points on the path that are as close as possible to the actual $\Phi$ and $\Psi$ values identified for each amino acid in the polypeptide chain. Thus, the process can be interpreted as a "step back" from the native structure of the chain in the direction of its original structure (Figure 1.3(c)) Applying this algorithm to a nonredundant set of proteins (derived from the PDB) produced the results depicted in Figure 1.3. The distribution of probability along the elliptical path (with the starting point marked in Figure 1.3(b)) clearly



Published by Woodhead Publishing Limited, 2012

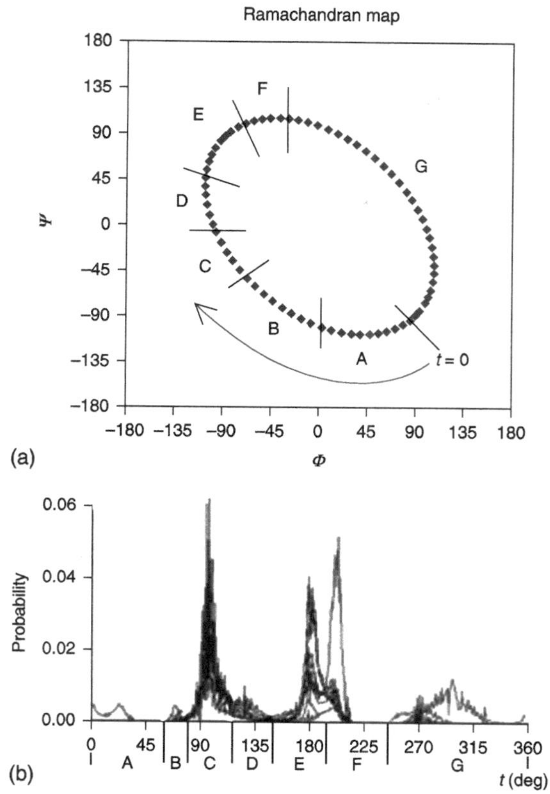


**Figure 1.3** Distribution of the likelihood of encountering a specific structure within the limited conformational subspace. (a) Direction of traversal of the elliptical path, corresponding to increasing angles *t* in the equation of the ellipse (horizontal axis – b). (b) Sample frequency distributions for 10 selected amino acids. The letters placed below the diagram correspond to local probability peaks. The same letters encode individual subsections of the elliptical path in part (c). (c) The "step-back" procedure, which transforms the native structure (expressed by appropriate angles $\Phi$ and $\Psi$) to its limited-subspace form ($\Phi_e$ and $\Psi_e$)

(*Continued*)



Published by Woodhead Publishing Limited, 2012

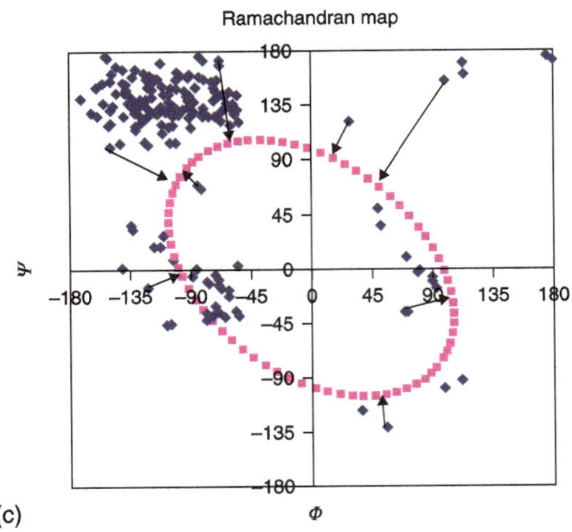


(c)

**Figure 1.3** *Continued*

points to the presence of seven local maxima, each of which has been tagged with a letter. Together, these letters constitute a structural alphabet that encodes a specific fragment of the path, reflecting its probability distribution peaks (Figure 1.3(b)).

Figure 1.4 shows a 3D representation of a sample protein (1ESL) for which the angles $\Phi$ and $\Psi$ have been calculated and subsequently transformed into corresponding angles $\Phi_e$ and $\Psi_e$. The secondary fragments in both of the structural forms are shown in Figure 1.5 to reveal the similarity of the ellipse-based structural form to the final form.

The structural codes are in a well-defined relation to secondary structural forms as follows: C, right-handed $\alpha$-helix; E and F, $\beta$-structural forms; G, left-handed $\alpha$-helix. The structural codes A, B, and D represent the unstructured forms known as "random coils."



Published by Woodhead Publishing Limited, 2012

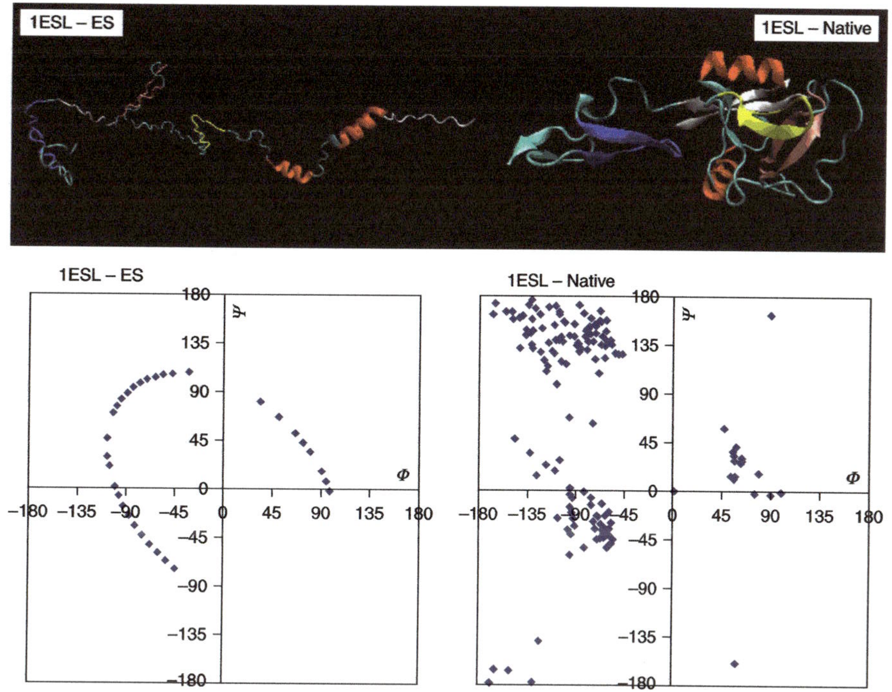


**Figure 1.4** Sample protein 1ESL for early-stage intermediate model. Top left, ES structure; top right, native structure. The lower row shows the Ramachandran maps for the ES intermediate (left) and the native structure (right). The secondary structural fragments are colored in both structural versions. The program VMD was used for visualization

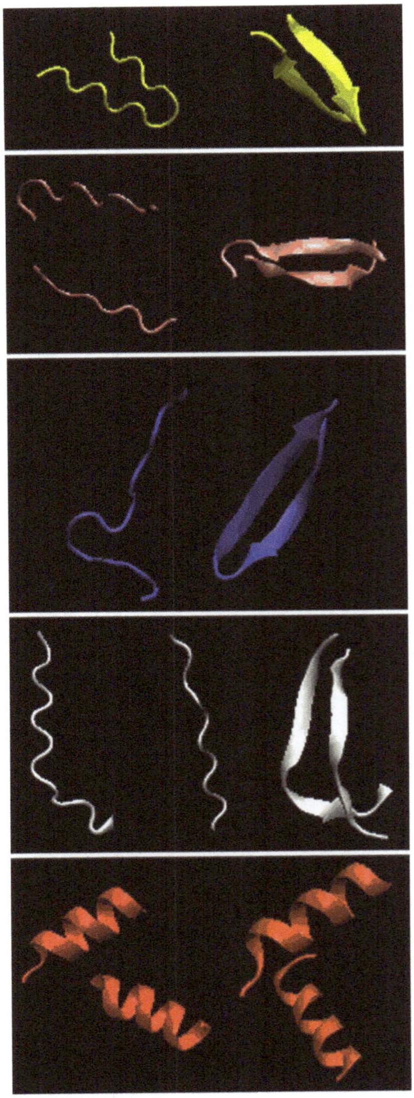

**Figure 1.5** Secondary-structure fragments in the ES form (left) and in the native form (right) of the protein 1ESL. Colors as in Figure 1.4. Note that the distances between the white and red fragments in the ES form are less than those in the complete structure shown in Figure 1.4. The program VMD was used for visualization



Published by Woodhead Publishing Limited, 2012

## 1.3 Contingency table

Introducing a simple structural encoding schema for the ES intermediate makes it possible to prepare a contingency table that expresses a relation between structural motifs and amino acid sequences.[5,6] For the purposes of this operation, the tetrapeptide was selected as a basic structural unit. The reason for this choice was twofold: the tetrapeptide chain is the shortest one for which the structure can be assigned to a specific secondary conformation ($\alpha$-helix, $\beta$-sheet, $\beta$-turn, etc.) and, moreover, the number of potential permutations remains manageable, enabling the preparation and managing of a reasonably sized table.

Given a full set of 20 amino acids, the number of potential tetrapeptides is 160 000. Each sequence of four peptides translates into a separate column in the contingency table. Since there are seven distinct structural codes, the number of potential combinations is 2401. The resulting table is quite large, but still compact enough to permit useful computations. Each cell in the table lists the probability of encountering a specific tetrapeptide with a specific combination of structural codes (the aggregate probability $p$ for all columns and all rows being equal to 1). In addition, the probability of occurrence of a given sequence for a single set of codes (with an aggregate probability equal to 1 within each row) and the probability of occurrence of a given set of codes for a single sequence (with an aggregate probability equal to 1 within each column) are also listed.

The individual cells list the following probability values ($p_{ij}^t$, $p_{ij}^c$, $p_{ij}^r$):

$$p_{ij}^t = \frac{n_{ij}}{N^t} \qquad [1.3]$$



Published by Woodhead Publishing Limited, 2012

$$p_{ij}^{c} = \frac{n_{ij}}{N_j^c} \qquad [1.4]$$

$$p_{ij}^{r} = \frac{n_{ij}}{N_i^r} \qquad [1.5]$$

where $i$ denotes the successive structures (rows), $j$ denotes the successive sequences (columns), $n_{ij}$ is the number of tetrapeptides representing sequence $i$ and structural motif $j$, $N_j^c$ and $N_i^r$ are the numbers of tetrapeptides representing sequence $i$ and structural motif $j$, respectively, and $N^t$ is the total number of cells in the table. Owing to the large size of the contingency table, the relatively small values of $p_{ij}^t$, $p_{ij}^c$, and $p_{ij}^r$ are expressed on a logarithmic scale.

Our analysis of the mutual interdependence of structures and sequences (based on the contingency table) exploits Shannon's theory of information.[7] This very universal and highly practical theory claims that the amount of information $I_i$ carried by an event $i$ with probability of occurrence $p_i$ equals

$$I_i = -\log_2 p_i \text{ (bits)} \qquad [1.6]$$

If several outcomes are possible, the corresponding information entropy $SE$ is given by

$$SE = -\sum_{i=1}^{N} p_i \log_2 p_i \text{ (bits)} \qquad [1.7]$$

where $N$ is the number of possible outcomes, with corresponding probabilities $p_i$.

The value of $SE$ can also be interpreted as the degree of uncertainty involved in predicting the outcome of an event. $SE$ reaches its peak value, $\overset{\max}{SE}$, if all potential outcomes are equally probable. In contrast, if only one outcome is possible, $SE$ becomes 0. The greater the difference between $\overset{\max}{SE}$ and $SE$, the more deterministic the event. High values of $\overset{\max}{SE} - SE$

calculated for cells of the contingency table may indicate a tight causative linkage between a given sequence and the corresponding structural motif. The value of $SE^r$ (calculated for each table row) can be interpreted as the preferred structural motif for a given sequence. Similarly, it is possible to establish preferred sequences for each structural motif, by assigning a dedicated value $SE^c$ to each table column, according to the following formula:

$$SE_j^c = -\sum_{i=1}^{N_j^0} p_{ij}^c \log_2 p_{ij}^c \qquad [1.8]$$

where $SE_j^c$ is the information entropy of the $j$-th column (sequence), $i$ is the row index, and $N_j^0$ denotes the number of nonempty cells in column $j$. $p_{ij}^c$ can be derived from Eq. 1.10 below. $SE_j^c$ makes a highly useful estimate of the uncertainty involved in predicting the structural counterpart of sequence $j$. Small values of $SE_j^c$ mean that the likelihood of making a correct prediction is high.

Since the number of nonempty cells varies, and $SE$ (as well as $\overset{\max}{SE}$) depends on the number of components, it is useful to introduce the peak entropy value $\overset{\max}{SE_j^c}$ for each sequence, which depends on the number of possible solutions, i.e., nonempty cells. $\overset{\max}{SE}$ is a quantitative measure of the degree of uncertainty in a situation where all outcomes are equally probable – thus, for the $j$-th column (sequence), $\overset{\max}{SE_j^c}$ can be calculated as

$$\overset{\max}{SE_j^c} = -\sum_{i=1}^{N_j^0} \overset{\max}{p_{ij}^c} \log_2 \overset{\max}{p_{ij}^c} \qquad [1.9]$$

where $i$ denotes a structural motif, $N_j^0$ is the number of nonempty cells in the $j$-th column, and $\overset{\max}{p_{ij}^c}$ is the corresponding probability, assuming that all structural motifs are equally well represented (as required by $\overset{\max}{SE}$). For each nonempty cell ($i = 1, \ldots, N_j^0$) in the $j$-th column, the value of $\overset{\max}{p_{ij}^c}$ is given by



Published by Woodhead Publishing Limited, 2012

$$\overset{\max}{p}{}_{ij}^{c} = \frac{1}{N_j^0} \qquad [1.10]$$

The values of $SE$ and $\overset{\max}{SE}$ tell us how far we are from total randomness (the worst-case scenario) – in other words, $\Delta SE^c$ expresses the degree of mutual interdependence between the sequences and structural motifs. Accordingly, each sequence is characterized by

$$\Delta SE_j^c = \overset{\max}{SE}{}_j^c - SE_j^c \qquad [1.11]$$

Similar calculations have been performed for all rows of the contingency table. Each possible structure has been parameterized with the values of $SE_i^r$, $\overset{\max}{SE}{}_i^r$, and $\Delta SE_i^r$.

According to the theory presented here, $\overset{\max}{SE}$ determines the peak entropy when all nonempty cells are ascribed equal probability values. Thus, for $SE_j^c = \overset{\max}{SE}{}_j^c$, all "legal" structures are equally likely to correspond to a given sequence, and the prediction boils down to a random guess. However, as $\overset{\max}{SE} - SE$ increases, so do the chances of making a correct prediction. This correlation justifies selecting $\Delta SE$ as a universal measure of structural/sequential interdependence in the contingency table. Results are presented in Figures 1.6 and 1.7, each limited to the 10 highest values of $\Delta SE$ (the strongest interdependence). In order to differentiate between structural codes and sequences, the former are italicized, and the latter are printed in boldface.

Figure 1.6 lists the 10 structural motifs with the highest probability of being correctly matched to tetrapeptide sequences. The results presented here may appear surprising: it turns out that the most accurate predictions can be made for highly unordered loop-like structures (note the poor representation of highly ordered *C* and *E* motifs). Evidently, whenever an unordered motif is called for in the tertiary



Published by Woodhead Publishing Limited, 2012

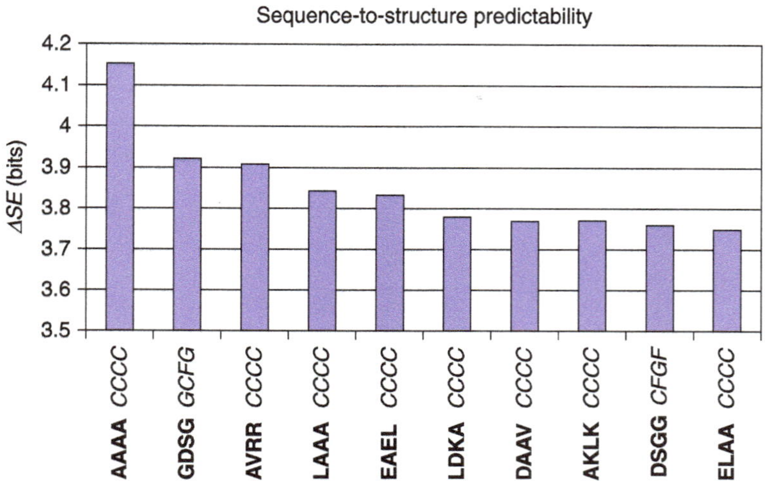


**Figure 1.6** **Sequence-to-structure dependence (i.e., the structural variability corresponding to a given sequence) expressed as the difference $\Delta SE$ between the observed information entropy $SE$ and its theoretical peak $\overset{max}{SE}$. Only the 10 sequences with the highest $\Delta SE$ values are listed. Sequence, bottom row (bold); structural codes, upper row (italic)**

structure of a protein, there are only a few amino acid sequences that can represent it.

A quick search for all nonzero $p$ values revealed that the database contains 2397 distinct combinations of structural codes (out of 2401 possible combinations) and 146940 distinct tetrapeptide sequences (out of 160000 possible combinations). Further analysis indicates that the maximum number of distinct structural motifs attributed to a single sequence is 144 (for the tetrapeptide **GSAA**). Conversely, the maximum number of distinct sequences corresponding to a single structural motif is 90 587 (*CCCC*, an $\alpha$-helix) and 47809 (*EEEE*, a $\beta$-sheet). The following four structures are not represented at all: *ABAB*, *ABBD*, *ABFB*, and *DBAB*.



Published by Woodhead Publishing Limited, 2012

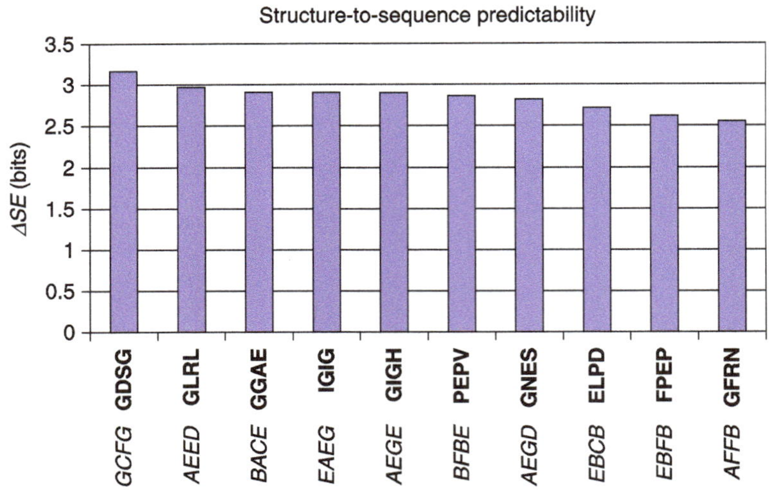


**Figure 1.7** **Structure-to-sequence dependence (i.e., the variability of the sequence for a given structural motif) expressed as the difference $\Delta SE$ between the observed information entropy $SE$ and its theoretical peak $\overset{max}{SE}$. Only the 10 structures with the highest $\Delta SE$ values are listed. Structural codes, bottom row (italic); sequence, upper row (bold)**

Given that *A* and *B* are the least frequently occurring local maxima (as seen in Figure 1.3(b)), this should come as no surprise. Searching for codes with the highest degree of determinism (defined as the lowest possible number of sequences for a given structure) revealed that they are primarily associated with loops and with characteristic amino acids such as GLY and PRO (though other amino acids – LEU, ILE, VAL, ARG, PHE, ASN, and ASP – also occur in this category).

It is interesting to note that the most typical structure for the ALA–ALA–ALA–ALA sequence is encoded as CCCC (which corresponds to an $\alpha$-helix) and, in general, all of the structures with the highest degree of determinism are helical in nature (as indicated by the presence of Cs in their corresponding structural codes).[5,6,8] The left-handed helix (identified by code *G*) is typically represented by tetrapeptides with GLY in their sequence.

A detailed analysis of the entire contingency table, together with a discussion of the quantitative properties of selected structural motifs and tetrapeptide sequences, can be found in Ref.9.

The contingency table was used to generate starting structures for free-energy optimization algorithms, which were assumed to predict the native forms of the proteins. In addition, a separate sequence/structure relationship study for early-stage folding intermediates was performed by way of a clustering analysis of the parameters *V* (angle) and *R* (radius of curvature). This study indicated the presence of 10 distinct clusters, i.e., frequently occurring structures with specific *V* and *R* values, eight of which correspond to common relations depicted in Figure 1.1(b). The remaining pair of clusters encapsulates "unexpected" relations and will be analyzed in Chapter 8.

## 1.4 In search of structural similarities

Translating the structure of a protein into a sequence of letter codes allows the structural similarity of early-stage folding intermediates to be studied. More specifically, this form of encoding enables the use of IT tools such as ClustalW, a software package dedicated to sequence similarity studies. As a result, searches for structural and sequential similarities become compatible in the sense that algorithms for the two purposes can share similar correspondence criteria.

Another way to discover structural similarities is by comparing the distributions of *V* and *R* (usually given as



Published by Woodhead Publishing Limited, 2012

ln($R$)). Low radii and tight angles indicate $\alpha$-helical structures, while high values correspond to $\beta$ motifs. The traditional classification of "random coil" gets its own representation, since the structural codes *A*, *B*, and *D* may successfully distinguish different forms of unstructured fragments.

An important advantage of this representation is that it enables researchers to visualize structures. Given a set of angles $\Phi$ and $\Psi$, this is possible only as a result of calculation. It is easy to transform a knowledge of the parameters *V* and *R* into a construction of the shape of the polypeptide.[10]

A case study was conducted on early-stage intermediates representing the serpine superfamily. The results indicate that the structures and, consequently, the biological properties of such proteins exhibit significant similarity (both structural and sequential), with particular focus on $\beta$-sheet motifs, which are known to play an important role in determining the biological function of serpines.[11] The structural alphabet was applied in a similarity study that used ClustalW to calculate a set of weight coefficients *W*, expressing the similarity at a particular position (and its corresponding protein fragment, used as a "window") with respect to both the amino acid sequences and the structural codes introduced in this chapter.

## 1.5 References


1. Alonso D.O., Daggett V. (1998) Molecular dynamics simulations of hydrophobic collapse of ubiquitin. *Protein Sci.* 7, 860–74.
2. Roterman I. (1995) The geometrical analysis of structural peptide backbone structure and its local deformations. *Biochimie* 77, 204–52.




Published by Woodhead Publishing Limited, 2012

3. Roterman I., Konieczny L. (1995) Geometrical analysis of structural changes in immunoglobin domains' transition from native to molten state. *Comput. Chem.* 19, 204–16.
4. Roterman I. (1995) Modelling the optimal simulation path in the peptide chain folding – studies based on geometry of alanine heptapeptide. *J. Theor. Biol.* 177, 283–8.
5. Han K.F., Baker D. (1996) Global properties of the mapping between local amino acid sequence and local structure in proteins. *Proc. Natl. Acad. Sci. USA* 93, 5814–18.
6. Lim V.I. (1974) Algorithms for prediction of alpha-helical and beta-structural regions in globular proteins. *J. Mol. Biol.* 88, 873–94.
7. Shannon C.E. (1948) A mathematical theory of communication. *Bell Syst. Tech. J.* 27, 379–423.
8. Chou P.Y., Fasman G.D. (1974) Conformational parameters for amino acids in helical, beta-sheet, and random coil regions calculated from proteins. *Biochemistry* 13, 211–22.
9. Bryliński M., Konieczny L., Czerwonko P., Jurkowski W., Roterman I. (2005) Early-stage folding in proteins (in silico) – sequence-to-structure relation. *J. Biomed. Biotechnol.* 2, 65–79.
10. Leluk J., Konieczny L., Roterman I. (2003) Search for structural similarity in proteins. *Bioinformatics* 19, 117–24.
11. Bryliński M., Konieczny L., Kononowicz A., Roterman I. (2008) Conservative secondary structure motifs already present in early-stage folding (*in silico*) as found in the serpine family. *J. Theor. Biol.* 251, 275–85.



Published by Woodhead Publishing Limited, 2012

## 1.6 Suggested reading

Benner S.A. (1989) Patterns of divergence in homologous proteins as indicators of tertiary and quaternary structure. *Adv. Enzyme Regul.* 28, 219–36.

Colubri A. (2004) Prediction of protein structure by simulating coarse-grained folding pathways: a preliminary report. *J. Biomol. Struct. Dyn.* 21, 625–38.

Fernandez A., Colubri A., Aqpigmanesi G., Burastero T. (2001) Coarse semiempirical solution to the protein folding problem. *Physica A* 293, 358–84.

Levitt M., Warshel A. (1975) Computer simulation of protein folding. *Nature* 253, 694–8.

Maxfield F.R., Scheraga H.A. (1979) Improvements in the prediction of protein backbone topography by reduction of statistical errors. *Biochemistry* 18, 697–704.

Mezei M. (2003) A novel fingerprint for the characterization of protein folds. *Protein Eng.* 16, 713–15.

Pappu R.V., Srinivasan R., Rose G.D. (2000) The Flory isolated-pair hypothesis is not valid for polypeptide chains: implications for protein folding. *Proc. Natl. Acad. Sci. USA* 97, 12565–70.

Ptitsyn O.B., Bychkova V.E., Uversky V.N. (1995) Kinetic and equilibrium folding intermediates. *Philos. Trans. R. Soc. Lond. B Biol. Sci.* 348(1323), 35–41.

Sosnick T.R., Berry R.S., Colubri A., Fernandez A. (2002) Distinguishing foldable proteins from nonfolders: when and how do they differ? *Proteins* 49, 15–23.

Zvelebil M.J., Barton G.J., Taylor W.R., Sternberg M.J. (1987) Prediction of protein secondary structure and active sites using the alignment of homologous sequences. *J. Mol. Biol.* 195, 957–61.



Published by Woodhead Publishing Limited, 2012

2

# The late-stage intermediate

*M. Banach, L. Konieczny, and I. Roterman-Konieczna*

**Abstract:** The second step in the protein folding process depends on interactions with external force fields and on an internal field that governs intramolecular interactions. The internal field comprises electrostatic, van der Waals, and torsion potentials (each of which can be optimized in a pairwise fashion), and the external field must take account of the presence of water. This results in the location of hydrophobic residues in the interior and simultaneous exposure of hydrophilic residues on the protein surface, where they are more likely to come into contact with water. According to our model, the structure of the protein's hydrophobic core can be described by a 3D Gaussian function. Thus, structural optimization depends on minimizing the differences between the theoretical (expected) distribution of hydrophobicity (which reaches its peak value in the core, while remaining close to 0 on the surface) and the observed distribution, which emerges as the polypeptide chain undergoes folding. The algorithm proposed here takes the two types of forces (internal and external) into account, and attempts to interleave internal optimization (i.e., atom–atom interactions) with external optimization, which directs hydrophobic residues towards the center of the protein.



Published by Woodhead Publishing Limited, 2012

**Key words:** oil drop model, fuzzy oil drop model, hydrophobicity, internal force field, external force field.

## 2.1 The "fuzzy oil drop" model

A crucial assumption underpinning the second stage of the folding process (the late stage, LS) is that this stage must take account of the presence of side chains (which were omitted in the ES step) and the external aqueous environment, both of which exert an influence on the final shape of the protein.

Internal interactions – traditionally expressed in the form of electrostatic interactions, van der Waals interactions, and torsion-potential forces – can be supplemented with hydrophobic interactions, affecting the polypeptide chain as a whole. Existing models tend to approach such interactions in a pairwise fashion: each pair of atoms (or side chains) is modeled in such a way that its corresponding interactions assume the lowest possible energy state for a given set of external conditions.

Our model assumes only classical pairwise optimization of the electrostatic, van der Waals, and torsion-potential forces. However, unlike classical approaches, it also involves the concept of modeling hydrophobic interactions with an external force field. Interactions – whether internal or external – determine the final arrangement of side chains and the effect exerted upon the protein molecule by its surrounding environment. Popular models approach this issue by, for example, tracking solvent-accessible surface areas,[1] measuring the so-called degree of exclusion,[2] assessing the depth at which hydrophobic residues are "buried" in the protein body,[3] or introducing an implicit representation of the aqueous solvent.[4] The model presented here takes into account the commonly known rule that the stabilization of



Published by Woodhead Publishing Limited, 2012

the protein's tertiary structure depends primarily upon hydrophobic interactions.[5]

As hydrophobic interactions stabilize the molecule as a whole, they cannot be approached in a pairwise fashion. Rather, a formal mechanism needs to be developed to represent the global influence of the environment on the protein as a whole. A 3D Gaussian function seems to be a good candidate for use in such a mechanism. This function makes it possible to calculate the hydrophobicity at arbitrary points in the three-dimensional structure of the protein. The 3D Gaussian function mimics a "fuzzy oil drop" with the highest hydrophobicity in the center and low hydrophobicity on the surface.

The 3D Gaussian function is given by

$$\tilde{H}t_j = \frac{1}{\tilde{H}t_{\text{sum}}} \exp\left(\frac{-\left(x_j - \bar{x}\right)^2}{2\sigma_x^2}\right) \exp\left(\frac{-\left(y_j - \bar{y}\right)^2}{2\sigma_y^2}\right) \exp\left(\frac{-\left(z_j - \bar{z}\right)^2}{2\sigma_z^2}\right) \qquad [2.1]$$

where $\bar{x}$, $\bar{y}$, $\bar{z}$ are the coordinates of the geometric center of the molecule (usually coinciding with the origin of the coordinate system) and, simultaneously, the center of the bounding ellipsoid. The size of the molecule (and of the "drop") is expressed by the triplet $\sigma_x$, $\sigma_y$, $\sigma_z$, provided that the molecule is oriented in such a way that the longest possible distance between two effective atoms coincides with an axis of the coordinate system. The coefficients $\sigma$ are given by one third of the longest distance between the effective atoms parallel to each axis. The value of the Gaussian function (at each point within the protein body) can be treated as a representation of the structure of the environment, with an idealized hydrophobicity distribution.

The 3D Gaussian function describes an ellipsoid with values increasing towards the center of the molecule (i.e., its hydrophobic core). This distribution of hydrophobicity,



Published by Woodhead Publishing Limited, 2012

which is expected in a stable protein, agrees with Kauzmann's model, which likens the protein to an "oil drop."[6] The presence of water causes hydrophobic residues to be located in the interior in much the same way as in a drop of oil, where contact between hydrophobic particles and the external environment is also minimized.[5] Hydrophilic residues, when exposed on the surface, may alter their conformation dynamically as a result of interaction with the solvent. Thus, we expect the final structure of the protein to be characterized by significant aggregation of hydrophobic residues in the core, together with exposure of hydrophilic residues on the surface.

At this point, it is necessary to find out whether our theoretical model corresponds to the actual conformation of residues in folded proteins. The observed distribution (hydrophobicity density) can be determined by applying an empirical function proposed by Levitt,[7]

$$\tilde{H}o_j = \frac{1}{\tilde{H}o_{\mathrm{sum}}} \sum_{i=1}^{N} (H_i^r + H_j^r) \times \begin{cases} \left[1 - \frac{1}{2}\left(7\left(\frac{r_{ij}}{c}\right)^2 - 9\left(\frac{r_{ij}}{c}\right)^4 + 5\left(\frac{r_{ij}}{c}\right)^6 - \left(\frac{r_{ij}}{c}\right)^8\right)\right] & \text{for } r_{ij} \le c \\ 0 \text{ for } r_{ij} > c \end{cases} \quad [2.2]$$

where $N$ is the number of amino acids in the protein (the number of grid points), $\tilde{H}_i^r$ expresses the hydrophobicity of the $i$-th residue according to an accepted hydrophobicity scale (which can be arbitrary – see, for example, Ref.8), $r_{ij}$ is the distance between the $i$-th and $j$-th interacting residues, and $c$ is a cutoff distance, which was assumed to be 9 Å in the original paper.[7] The values of $\tilde{H}o_j$ are standardized by division by the coefficient $\tilde{H}o_{\mathrm{sum}}$, which is the aggregate sum of hydrophobicity values attributed to all grid points.

In order to facilitate meaningful comparisons between the theoretical and actual distributions of hydrophobicity, the two results need first to be normalized by introducing a coefficient (the initial factor in the two formulae). Each result is divided by the sum of the hydrophobicity values assigned to all residues – thus, the aggregate hydrophobicity for the entire protein is always equal to 1. Normalization enables us to present results in the form of a profile expressing the hydrophobicity differences for individual residues in the polypeptide chain. In order to better illustrate this concept, let us focus on two case studies, described below.

## 2.2 Quantitative description of the hydrophobic core

Before the model can be applied to the simulation of protein folding, it needs to be checked to what extent real structures follow the assumptions of the model.

The distributions depicted in Figure 2.1 (the theoretical distribution $\tilde{H}_t$ and the observed distribution $\tilde{H}_o$) give an idea of the extent of the observed differences. In order to quantitatively assess the degree of correspondence between the two distributions, the concept of the divergence entropy proposed by Kullback and Leibler[9] can be applied. Originally meant as a measure of the separation between two distinct probability distributions, it can also be applied to other quantities (such as hydrophobicity distributions, which can be treated as the probability of the occurrence of hydrophobicity). It is given by

$$D_{KL}(p|p^0) = \sum_{i=1}^{N} p_i \log_2(p_i / p_i^0), \qquad [2.3]$$

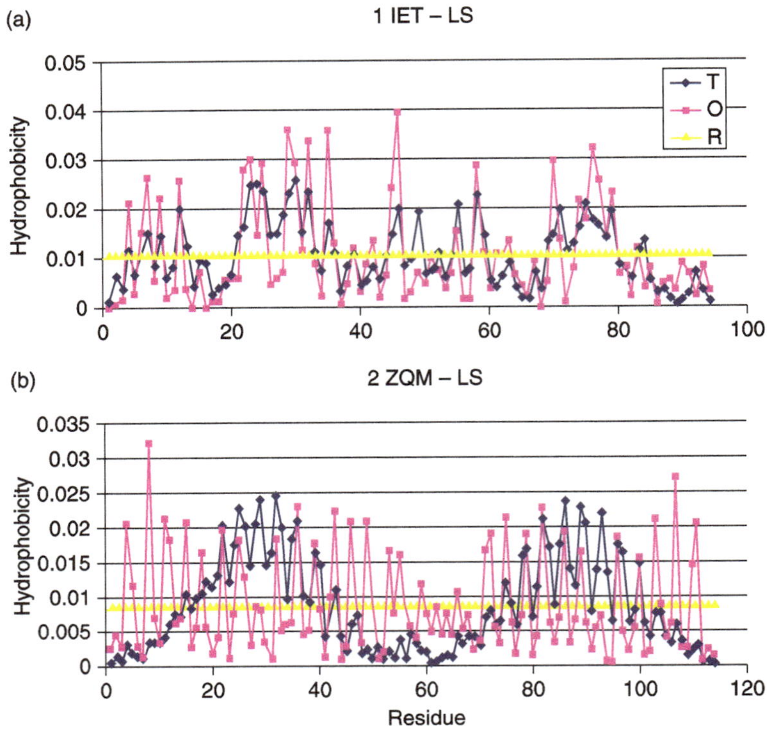


**Figure 2.1** **Sample proteins exhibiting good ((a), 1IET) and poor ((b), 2ZQM) agreement between theoretical and observed hydrophobicity distributions. T, theoretical distribution (dark rhombuses); O, observed distribution (pink squares); R, random distribution (yellow triangles)**

where $D_{KL}$ is the divergence entropy, $p$ is the probability of occurrence of a particular observed event (or the hydrophobicity), and $p^0$ is the reference (target) probability (or hydrophobicity) in the reference distribution. The index $i$ denotes a particular amino acid, and $N$ is the total number of amino acids in the polypeptide chain. In the simplified form that we have adopted for the case under consideration, the above formula can be written as



Published by Woodhead Publishing Limited, 2012

$$O/T = \sum_{i=1}^{N} O_i \log_2 O_i / T_i \qquad [2.4]$$

where $O_i$ is a measure of the observed hydrophobicity of the $i$-th residue and $T_i$ is the corresponding theoretical hydrophobicity. *O/T* expresses the theoretical-versus-observed divergence entropy $D_{KL}$.

Since $D_{KL}$ is a measure of entropy, its absolute value cannot be properly interpreted without a reference. In the formula presented here, the target is provided by the theoretical distribution of hydrophobicity for a given polypeptide chain. This distribution is not, however, our only option – a random distribution can also be used as a target distribution, where each residue is assigned a value of $1/N$ ($N$ being the number of residues in the polypeptide chain, and thus $1/N$ represents a perfectly random distribution of hydrophobicity). The word "random" is used here in the sense that any residue positioned at any point in the space in the body of the oil drop has the same probability, equal to $1/N$. In this way, the observed distribution can be compared with both a highly ordered 3D Gaussian distribution and a distribution in which all residues carry the same probability of being identified as hydrophobic.

For the random reference, the entropy formula assumes the following form:

$$O/R = \sum_{i=1}^{N} O_i \log_2 O_i / R_i \qquad [2.5]$$

where *R* is the random target distribution. The value of *O/T* (which represents the divergence between the observed and theoretical distributions) can be compared with the value of *O/R* (which corresponds to the observed and random distributions). Cases where $O/T < O/R$ are treated as



Published by Woodhead Publishing Limited, 2012

representing good agreement between the observed and theoretical distributions (the presence of a hydrophobic core with a high concentration of hydrophobicity in the center of the protein molecule), while $O/T > O/R$ indicates a more randomized distribution of the actual (observed) hydrophobicity, suggesting the absence of the expected organization of the hydrophobicity in the form of a hydrophobic core.

In proteins that satisfy $O/T < O/R$, a Gaussian-like structure can usually be discerned, with an internal core containing highly hydrophobic residues. The hydrophobicity decreases as we approach the molecular surface, becoming close to 0 on the surface itself (at a distance of $3\sigma$ from the center of the Gaussian ellipsoid).

For the protein depicted in Figure 2.1(a), $O/T = 0.271$ and $O/R = 0.586$. The difference between these values suggests a Gaussian-like conformation. In contrast, the protein depicted in Figure 2.1(b) has $O/T = 0.735$ and $O/R = 0.448$, indicating that its structure cannot be accurately described by a 3D Gaussian function.

Assessment of the protein 2ZQM, which consists of helical and $\beta$-structural fragments and is not globular (which also implies that it does not contain a hydrophobic core), suggests good accordance with the ES model. The parameters $V$ and $R$ calculated for this protein lie on a curve that expresses their expected correspondence (see Figure 1.1 and Ref. 10).[10] Thus, in the case of 2ZQM, the folding process can be said to have terminated at the intermediate early stage. Since this protein (a prefolded $\beta$ subunit from *Thermococcus* strain KS-1) is known to perform the role of a chaperone,[10,11] it must presumably fold unerringly, regardless of the external conditions – factors other than water.



Published by Woodhead Publishing Limited, 2012

## 2.3 Protein characteristics with respect to the hydrophobic core

An interesting question arises: does the existence of proteins that diverge from the Gaussian model disprove the hypothesis about tertiary-structure stabilization presented here? The answer is "no." Let us consider an "ideal" hydrophobic core, in perfect agreement with the 3D Gaussian function. Clearly, such a protein would be highly stable (assuming proper pairwise optimization of internal forces) and would remain in equilibrium with the surrounding aqueous environment. It would exhibit excellent solubility – which may be a desirable feature – but it would not interact with any medium other than water. In particular, it would be unable to form complexes and bind ligands. For this reason, one should not expect to find that the hydrophobic core will conform to theoretical predictions (i.e., a 3D Gaussian function) with perfect accuracy in all or even a majority of natural proteins.

Having analyzed a large set of proteins, we can identify (so far) two groups that exhibit particularly good agreement between the observed and theoretical hydrophobicity distributions. The first group comprises antifreeze proteins,[10] whose biological role (disruption of ice crystal formation) requires good solubility but does not involve the binding of ligands or formation of complexes (it does not require a specific binding site). In fact, their existence proves that there is a biological niche for proteins that must simply reside in an aqueous environment and do not require specific interactions with other molecules.

The second group consists of the so-called downhill (fast-folding) proteins, which undergo very rapid and reversible folding *in vitro*, suggesting that their conformation is



Published by Woodhead Publishing Limited, 2012

dependent solely on the surrounding environment. The influence of water on the native structure of fast-folding proteins manifests itself in the fact that hydrophobic residues are located in the interior, with simultaneous exposure of hydrophilic residues on the molecular surface.[12]

The characteristics of both of these groups support the validity of the model presented here and help explain why the need for intermolecular interactions (a common feature of proteins) limits the degree to which theoretical models reflect the actual distribution of hydrophobicity in most protein groups.

## 2.4 Simulation of late-stage folding

The generation of the late-stage structural form according to the "fuzzy oil drop" model proceeds as follows:

1. The ES geometric intermediate (see Chapter 1) is taken as the starting structure for the LS model.
2. The following operations are applied iteratively:

   *Step A*. The initial orientation of the molecule is adjusted, as described earlier in this chapter. A 3D Gaussian "drop" is then generated to encapsulate the molecule. A 3D grid system is constructed. Each grid point represents the expected hydrophobicity (calculated according to a 3D Gaussian function).

   *Step B*. Minimization of the internal free energy (expressed in terms of electrostatic, van der Waals, and torsion-potential forces) is performed assuming pairwise interactions.



Published by Woodhead Publishing Limited, 2012

*Step C*. The structure generated in step B is subjected to optimization of the hydrophobic interactions. The convergence criterion is the minimization of the difference between the observed and theoretical hydrophobicity distributions,

$$\Delta\tilde{H} = \sum_{i=1}^{N}(\tilde{H}t_i - \tilde{H}o_i)^2 \qquad [2.6]$$

$\Delta\tilde{H}$ values equal to 0 identify a structure whose hydrophobic core matches the 3D Gaussian function. The index $i$ denotes the position of the grid point characterized by $\tilde{H}t$ (the theoretical distribution in accordance with the 3D Gaussian function) and $\tilde{H}o$ (the distribution observed as a result of local hydrophobic interactions between residues).

*Step D*. The optimization procedure presented in step C upsets the results of the free-energy optimization performed earlier. For this reason, another round of free-energy optimization is necessary.[13]

*Step E*. The next step involves optimization of the hydrophobicity, performed in an external force field (a 3D Gaussian function) for a smaller "drop" (Figure 2.2).

*Step F*. Steps C, D, and E are performed iteratively. Each time, the size of the bounding drop is decreased (the final size of the drop is dependent on the length of the polypeptide chain and has been calculated for some proteins found in the PDB).[14]

As mentioned earlier, not all proteins follow the idealized hydrophobicity distribution. The reasons for these discrepancies, and their sources, will be discussed in Chapter 4.



Published by Woodhead Publishing Limited, 2012

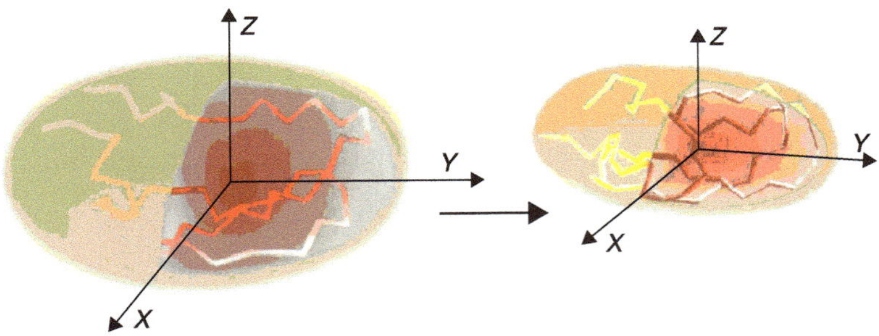


**Figure 2.2** **Model depiction of a 3D Gaussian function representing the external environment. Each step of the "fuzzy oil drop" algorithm reduces the volume of the bounding drop until a critical volume (depending on the length of the polypeptide chain) is reached. Shades of red represent variable hydrophobicity in the "fuzzy oil drop"**

## 2.5 References


1. Carugo O. (2000) Predicting residue solvent accessibility from protein sequence by considering the sequence environment. *Protein Eng. Des. Sel.* 13, 607–9.
2. Li J., Liu Q. (2009) "Double water exclusion": a hypothesis refining the O-ring theory for the hot spots at protein interfaces. *Bioinformatics* 25, 743–50.
3. Betts S., Haase-Pettingell C., Cook K., King J. (2004) Buried hydrophobic side-chains essential for the folding of the parallel beta-helix domains of the P22 tailspike. *Protein Sci.* 13(9), 2291–303.
4. Roux B., Simonson T. (1999) Implicit solvent models. *Biophys. Chem.* 78(1–2), 1–20.
5. Konieczny L., Brylinski M., Roterman I. (2006) Gauss-function-based model of hydrophobicity density in proteins. *In Silico Biol.* 6, 5–22.
6. Kauzmann W. (1959) Some factors in the interpretation of protein denaturation. *Adv. Protein Chem.* 14, 1–63.




Published by Woodhead Publishing Limited, 2012

7. Levitt M. A simplified representation of protein conformations for rapid simulation of protein folding. *J Mol. Biol.* 104 (1976) 59–107.
8. Brylinski M., Konieczny L., Roterman I. (2007) Is the protein folding an aim-oriented process? Human haemoglobin as example. *Int. J. Bioinform. Res. Appl.* 3(2), 234–60.
9. Nalewajski R.F. (2006) *Information Theory of Molecular Systems*. Amsterdam: Elsevier.
10. Banach M., Prymula K., Jurkowski W., Konieczny L., Roterman I. (2012) Fuzzy oil drop model to interpret the structure of antifreeze proteins and their mutants. *J. Mol. Model.* 18(1), 229–37.
11. Kida H., Sugano Y., Iizuka R., Fujihashi M., Yohda M., Miki K. (2008) Structural and molecular characterization of the prefold in beta subunit from *Thermococcus* strain KS–1. *J. Mol. Biol.* 383, 465–74.
12. Roterman I., Konieczny L., Jurkowski W., Prymula K., Banach M. (2011) Two-intermediate model to characterize the structure of fast-folding proteins. *J. Theor. Biol.* 283(1), 60–70.
13. Roterman I., Konieczny L., Banach M., Jurkowski W. (2011) Intermediates in the protein folding process: a computational model. *Int. J. Mol. Sci.* 12(8), 4850–60.
14. Brylinski M., Konieczny L., Roterman I. (2006) Fuzzy-oil-drop hydrophobic force field – a model to represent late-stage folding (in silico) of lysozyme. *J. Biomol. Struct. Dyn.* 23(5), 519–28.

## 2.6 Suggested reading

Adhikari A.N., Peng J., Wilde M., Xu J., Freed K.F., Sosnick T.R. (2012) Modeling large regions in proteins:



Published by Woodhead Publishing Limited, 2012

applications to loops, termini, and folding. *Protein Sci.* 21(1), 107–21.

Baurin N., Vangrevelinghe E., Morin-Allory L., Merour J.Y., Renard P., et al. (2000) 3D-QSAR CoMFA study on imidazolinergic I(2) ligands: a significant model through a combined exploration of structural diversity and methodology. *J. Med. Chem.* 43, 1109–22.

Bork P., Dandekar T., Diaz-Lazcoz Y., Eisenhaber F., Huynen M., Yuan Y. (1998) Predicting function: from genes to genomes and back. *J. Mol. Biol.* 283, 707–25.

Bouziane H., Messabih B., Chouarfia A. (2011) Profiles and majority voting-based ensemble method for protein secondary structure prediction. *Evol. Bioinform. Online* 7, 171–89.

Burley S.K., Almo S.C., Bonanno J.B., Capel M., Chance M.R., et al. (1999) Structural genomics: beyond the Human Genome Project. *Nat. Genet.* 23, 151–7.

Chance M.R., Bresnick A.R., Burley S.K., Jiang J.S., Lima C.D., et al. (2002) Structural genomics: a pipeline for providing structures for the biologist. *Protein Sci.* 11, 723–38.

Chandonia J.M., Earnest T.N., Brenner S.E. (2004) Structural genomics and structural biology: compare and contrast. *Genome Biol.* 5, 343.

Cramer R.D. 3rd, Patterson D.E., Bunce J.D. (1989) Recent advances in comparative molecular field analysis (CoMFA). *Prog. Clin. Biol. Res.* 291, 161–5.

Danielson M.L., Lill M.A. (2012) Predicting flexible loop regions that interact with ligands: the challenge of accurate scoring. *Proteins* 80(1), 246–60.

Devos D., Valencia A. (2000) Practical limits of function prediction. *Proteins* 41, 98–107.

Garcia P., Bruix M., Rico M., Ciofi-Baffoni S., Banci L., et al. (2005) Effects of heme on the structure of the denatured



Published by Woodhead Publishing Limited, 2012

state and folding kinetics of cytochrome b562. *J. Mol. Biol.* 346, 331–44.

Ghalwash M.F., Dunker A.K., Obradovi Z. (2012) Uncertainty analysis in protein disorder prediction. *Mol. Biosyst.* 8(1), 381–91.

Goulding C.W., Apostol M., Anderson D.H., Gill H.S., Smith C.V., et al. (2002) The TB structural genomics consortium: providing a structural foundation for drug discovery. *Curr. Drug Targets Infect. Disord.* 2, 121–41.

Hegyi H., Gerstein M. (1999) The relationship between protein structure and function: a comprehensive survey with application to the yeast genome. *J. Mol. Biol.* 288, 147–64.

Jones S., Thornton J.M. (1997) Analysis of protein–protein interaction sites using surface patches. *J. Mol. Biol.* 272, 121–32.

Kanaya S., Oobatake M., Liu Y. (1996) Thermal stability of *Escherichia coli* ribonuclease HI and its active site mutants in the presence and absence of the $Mg^{2+}$ ion. Proposal of a novel catalytic role for Glu48. *J. Biol. Chem.* 271, 32729–36.

Lamb M.L., Burdick K.W., Toba S., Young M.M., Skillman A.G., et al. (2001) Design, docking, and evaluation of multiple libraries against multiple targets. *Proteins* 42, 296–318.

Li D., Li T., Cong P., Xiong W., Sun J. (2012) A novel structural position-specific scoring matrix for the prediction of protein secondary structures. *Bioinformatics* 28(1), 32–9.

Liang J., Edelsbrunner H., Woodward C. (1998) Anatomy of protein pockets and cavities: measurement of binding site geometry and implications for ligand design. *Protein Sci.* 7, 1884–97.

McGuffin L.J., Roche D.B. (2011) Automated tertiary structure prediction with accurate local model quality



Published by Woodhead Publishing Limited, 2012

assessment using the IntFOLD-TS method. *Proteins* 79 Suppl. 10, 137–46.

Meiering E.M., Serrano L., Fersht A.R. (1992) Effect of active site residues in barnase on activity and stability. *J. Mol. Biol.* 225, 585–9.

Nagata K., Randall A., Baldi P. (2012) SIDEpro: a novel machine learning approach for the fast and accurate prediction of side-chain conformations. *Proteins* 80(1), 142–53.

Ondrechen M.J., Clifton J.G., Ringe D. (2001) THEMATICS: a simple computational predictor of enzyme function from structure. *Proc. Natl. Acad. Sci. USA* 98, 12473–8.

Oshiro C.M., Kuntz I.D., Dixon J.S. (1995) Flexible ligand docking using a genetic algorithm. *J. Comput. Aided Mol. Des.* 9, 113–30.

Polanski J., Walczak B. (2000) The comparative molecular surface analysis (COMSA): a novel tool for molecular design. *Comput. Chem.* 24, 615–25.

Pritchard-Bell A., Shell M.S. (2011) Smoothing protein energy landscapes by integrating folding models with structure prediction. *Biophys J.* 101(9), 2251–9.

Shoichet B.K., Baase W.A., Kuroki R., Matthews B.W. (1995) A relationship between protein stability and protein function. *Proc. Natl. Acad. Sci. USA* 92, 452–6.

Sippl W. (2002) Development of biologically active compounds by combining 3D QSAR and structure-based design methods. *J. Comput. Aided Mol. Des.* 16, 825–30.

Skolnick J., Fetrow J.S. (2000) From genes to protein structure and function: novel applications of computational approaches in the genomic era. *Trends Biotechnol.* 18, 34–9.

Toon A., Williams G. (2012) A dynamical approach to contact distance based protein structure determination. *J. Mol. Graph. Model.* 32, 75–81.



Published by Woodhead Publishing Limited, 2012

Wallace A.C., Borkakoti N., Thornton J.M. (1997) TESS: a geometric hashing algorithm for deriving 3D coordinate templates for searching structural databases. Application to enzyme active sites. *Protein Sci.* 6, 2308–23.

Wilson C.A., Kreychman J., Gerstein M. (2000) Assessing annotation transfer for genomics: quantifying the relations between protein sequence, structure and function through traditional and probabilistic scores. *J. Mol. Biol.* 297, 233–49.

Xu D., Zhang Y. (2011) Improving the physical realism and structural accuracy of protein models by a two-step atomic-level energy minimization. *Biophys. J.* 101(10), 2525–34.

Zhang J., Liu Z.P., Jones T.A., Gierasch L.M., Sambrook J.F. (1992) Mutating the charged residues in the binding pocket of cellular retinoic acid-binding protein simultaneously reduces its binding affinity to retinoic acid and increases its thermostability. *Proteins* 13, 87–99.

Zhang J., He Z., Wang Q., Barz B., Kosztin I., et al. (2012) Prediction of protein tertiary structures using MUFOLD. *Methods Mol. Biol.* 815, 3–13.

Zvelebil M.J., Sternberg M.J. (1988) Analysis and prediction of the location of catalytic residues in enzymes. *Protein Eng.* 2, 127–38.



Published by Woodhead Publishing Limited, 2012

# 3

# Structural information involved in the interpretation of the stepwise protein folding process

*P. Alejster, W. Jurkowski, and I. Roterman-Konieczna*

**Abstract:** Calculating the quantity of information present in each step of the protein folding process suggests that the multistep approach requires less information than the one-step model. Quantitative analysis reveals that the amino acids present in the polypeptide chain do not carry enough information to accurately predict the values of the angles $\Phi$ and $\Psi$ in folded proteins. This conclusion results from comparing the amount of information carried by amino acids with the quantity of information necessary to determine $\Phi$ and $\Psi$, taking the complete Ramachandran map as the conformational space. It is shown that the two-step model (comprising two stages, the ES and LS) requires less information, owing to the fact that the final predictions of the angles $\Phi$ and $\Psi$ can be based on a preexisting ES structure. Analysis based on information theory points to particular zones of the Ramachandran map that appear to play an important role in the context of protein structure prediction.



Published by Woodhead Publishing Limited, 2012

**Key words:** information theory, quantity of information, information entropy, bit, Ramachandran map, one-step versus two-step model.

## 3.1 Balancing the quantity of information in the amino acid sequence and the early-stage intermediate

Let us assume that the model described previously remains valid, i.e.,

$$U \rightarrow ES \rightarrow LS \rightarrow N$$

An important question arises: how to formally validate the introduction of two intermediate stages? In order to provide an answer, let us refer to the basic concepts of information theory. The quantity of information carried by an event whose probability of occurrence is $p$ can be defined as follows[1]:

$$I_i = -\log_2 p_i \qquad [3.1]$$

Specifically, $I_i$ denotes the quantity of information (in bits) carried by an elementary event with probability $p_i$. The information entropy, interpreted as a weighted-average quantity of information for a set of outcomes (each of which represents a possible realization of an elementary event), is given by the following formula:

$$SE = -\sum_{i=1}^{n} p_i \log_2 p_i \qquad [3.2]$$

where $n$ denotes the total number of outcomes of the event analyzed, each outcome having a probability $p_i$.

In the context of the protein folding process, the input data is limited to a sequence of amino acids. Using Eq. 3.1

and the frequency of occurrence of each amino acid in natural proteins ($p_i$), the quantity of information carried by each amino acid can be accurately calculated. These quantities, derived for each amino acid on the basis of a nonredundant set of PDB data, are listed in Figure 3.1. (Note that if all amino acids were equally likely to occur, i.e., if $p_i$ were equal to 1/20 for all $i$, the quantity of information per amino acid would amount to 4.322 bits.)

The expected output is a structural representation of the protein, encoded as a sequence of angles $\Phi$ and $\Psi$ for each amino acid. An interesting question arises: how much information is required to correctly predict the values of each angle?

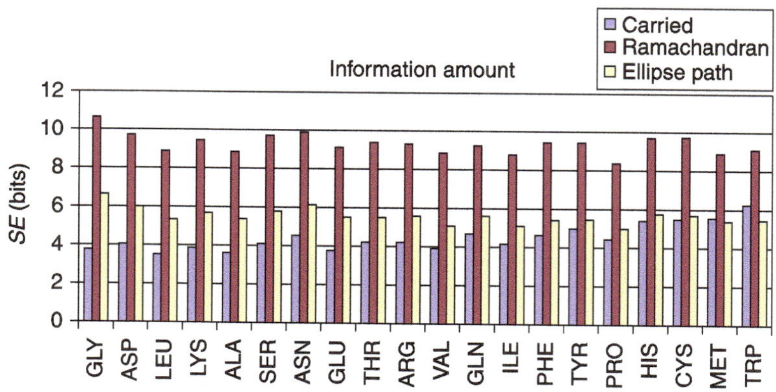


**Figure 3.1** **Quantity of information (bits) carried by each amino acid, taking into account its frequency of occurrence (results obtained for a nonredundant set of proteins using the BLAST software package with the nonredundancy criterion set to $p = 10^{-7}$, shown by blue bars). The *SE* values (Eq. 3.2) for the entire Ramachandran map with a 5° step size are shown by brown bars, and the corresponding information entropy for the elliptical subspace previously discussed is shown by white bars**



Published by Woodhead Publishing Limited, 2012

Each point on the Ramachandran map corresponds to a specific pair of angles $\Phi$ and $\Psi$ and can be assigned a specific probability of occurrence for each amino acid (see Figure 3.2). As can be expected, high probabilities are associated with regions that correspond to known secondary structural motifs. In order to study this phenomenon, the probability of occurrence of each pair of angles $\Phi$ and $\Psi$ was calculated at intervals of 1, 5, and 10°. This calculation was performed for a nonredundant selection of proteins from the PDB database. The quantity of information required to formulate a prediction for each pair of angles (for a given grid density), i.e., the corresponding entropy (the average quantity of information represented by a given conformation of a specific amino acid), is shown in Figure 3.1 (additional lists can be found in Refs 2 and 3).

Comparing the values presented here indicates that the expected solution requires more information than can be provided by the input chain. If, however, the search is restricted to the elliptical path (the limited conformational subspace), it becomes possible to model the early-stage

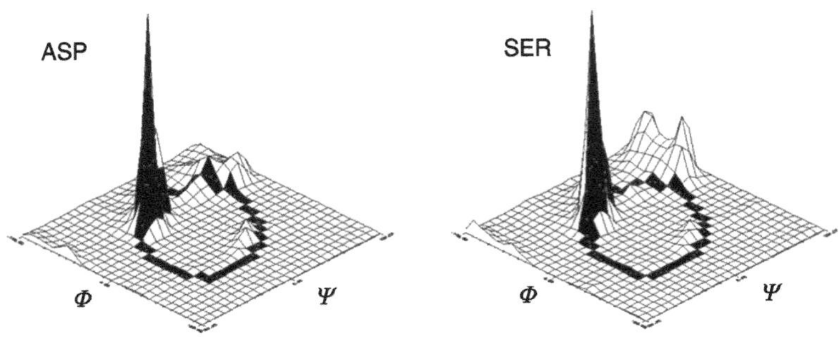


**Figure 3.2** **Distributions of probability for two sample amino acids. The elliptical path (marked in black) shows that, in both cases, the peaks in the distribution (secondary structures) overlap with or touch the limited conformational subspace**



Published by Woodhead Publishing Limited, 2012

intermediate with an accuracy of 5° per amino acid. While certain amino acids still exhibit information deficiencies, others compensate for this by providing excess information. The net result is an information balance between the input chain and the resulting ES intermediate structure (accurate to within 5° per angle) – but note that the same does not hold for the final structure of the protein.

One can therefore conclude that the information carried by the input sequence is sufficient to model the ES intermediate, but not sufficient to progress beyond that point. This observation can be treated as justifying the need for an ES intermediate.

## 3.2 Zones on the Ramachandran map

Introducing structural encoding for the early-stage intermediate (see Figure 1.3(b)) results in a set of so-called "zones." Each zone is a fragment of the Ramachandran map, within which all points representing the angles $\Phi$ and $\Psi$ correspond to a specific fragment of the elliptical path associated with a local maximum. Figure 3.3 depicts the distribution of these zones.

It is easy to recognize that code C spans the $\alpha$-helix region, codes E and F correspond to $\beta$-sheet motifs, and code G represents a left-handed helix. The remaining codes (A, B, and D) represent structures commonly referred to as random coils (which can be quite varied, particularly with respect to their radii of curvature $R$ and their V-angles).

Some additional properties of the zones are presented in Table 3.1. *SE* values were calculated for each zone individually to show the accuracy of the predictions depending on the zone in question (note that the *SE* values reflect the size of



Published by Woodhead Publishing Limited, 2012

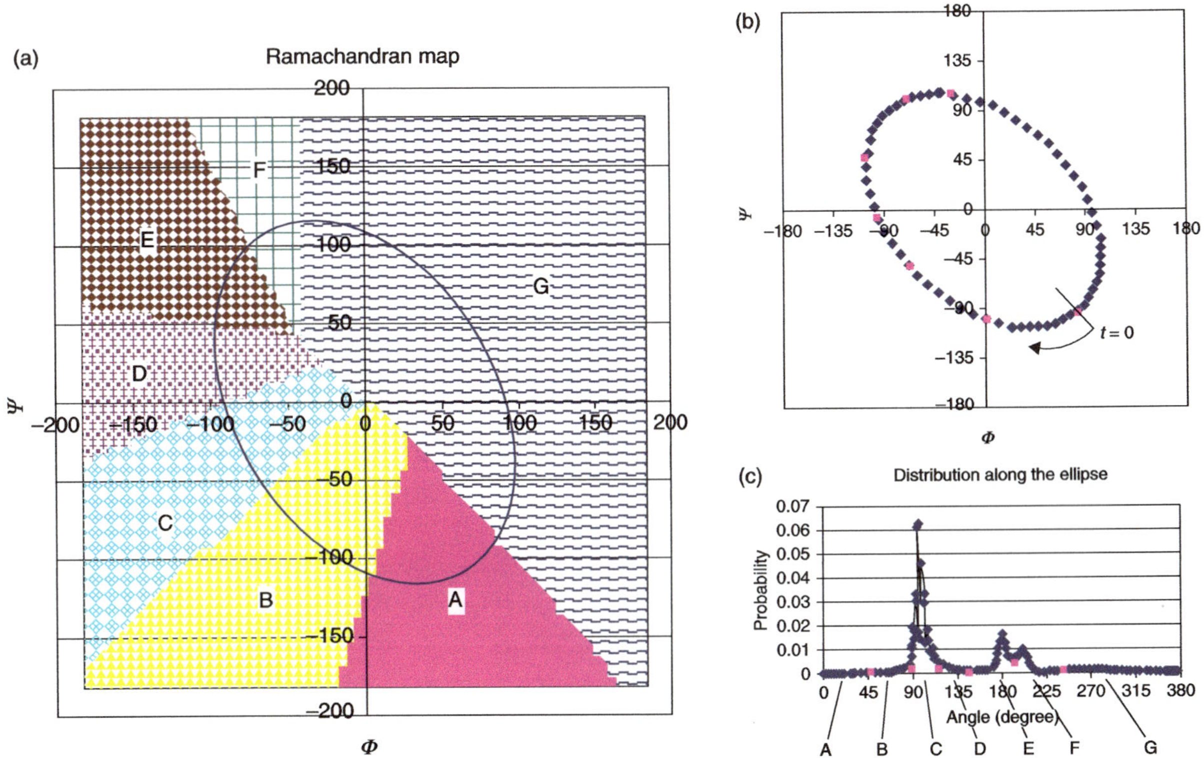


**Figure 3.3** (a) Zones of the Ramachandran map associated with specific local maxima of the probability distribution profile shown in (c). For the sake of clarity, the elliptical conformational subspace (b) is also illustrated, indicating the direction of traversal that corresponds to the profile presented in (c), as well as the boundaries between zones

**Table 3.1** ***SE* values (bits) for zones, for all amino acids**

| Amino acids | Zones | | | | | | |
|---|---|---|---|---|---|---|---|
| | A | B | C | D | E | F | G |
| ALA | 8.345 | 6.321 | 5.951 | 6.980 | 8.437 | 6.920 | 9.596 |
| ASN | 8.345 | 7.349 | 6.943 | 7.201 | 8.584 | 7.094 | 8.860 |
| ASP | 8.258 | 7.272 | 6.791 | 7.200 | 8.543 | 7.094 | 8.860 |
| ARG | 8.237 | 6.590 | 6.305 | 7.225 | 8.418 | 6.999 | 9.131 |
| CYS | 8.343 | 7.511 | 3.699 | 7.336 | 2.669 | 7.104 | 9.421 |
| GLN | 8.396 | 6.765 | 6.266 | 7.004 | 3.842 | 6.962 | 9.148 |
| GLU | 8.435 | 6.446 | 1.941 | 7.074 | 8.462 | 7.098 | 9.503 |
| GLY | 8.196 | 7.340 | 7.711 | 7.405 | 8.503 | 7.108 | 9.235 |
| HIS | 7.750 | 6.833 | 6.267 | 7.757 | 8.513 | 6.970 | 8.680 |
| ILE | 8.524 | 6.038 | 6.148 | 7.083 | 7.803 | 6.796 | 9.910 |
| LEU | 8.361 | 6.183 | 6.148 | 7.181 | 7.996 | 6.915 | 9.482 |
| LYS | 8.412 | 6.756 | 6.474 | 7.217 | 8.466 | 7.202 | 9.046 |
| MET | 7.984 | 6.343 | 6.083 | 6.967 | 8.315 | 7.028 | 8.980 |
| PHE | 7.810 | 6.306 | 6.354 | 7.323 | 8.254 | 7.043 | 8.823 |
| PRO | 9.000 | 7.052 | 6.434 | 6.718 | 4.972 | 7.256 | 11.177 |
| SER | 8.529 | 7.540 | 6.668 | 7.421 | 8.360 | 6.985 | 9.667 |
| THR | 8.518 | 7.334 | 6.740 | 7.327 | 8.044 | 6.948 | 10.014 |
| TRP | 7.306 | 6.409 | 6.348 | 7.065 | 8.127 | 6.845 | 8.838 |
| TYR | 8.041 | 6.467 | 6.431 | 7.220 | 8.216 | 6.892 | 8.937 |
| VAL | 8.701 | 7.566 | 6.463 | 7.262 | 8.009 | 6.744 | 9.667 |

each zone). These values also express the amount of information necessary to predict the angles $\Phi_z$ and $\Psi_z$ in the final structure, taking into account the conditional probability $p_i$ for a defined zone $z$ (i.e., the probability of encountering particular angles $\Phi_e$ and $\Psi_e$ within the fragment of the elliptical path recognized in the ES step).

Zone C appears to be the easiest one to make predictions for, owing to its relatively small size and high concentrations



Published by Woodhead Publishing Limited, 2012

of the angles $\Phi$ and $\Psi$ in the $\alpha$-helix region. VAL seems easy to attribute to zone F, unlike zone B, where its *SE* value is the highest. Zone A seems to enable accurate predictions for TRP, while zone D is likewise "compatible" with PRO. The predictability of HIS is similar to that of VAL owing to its high preference for zone G and low preference for zone D.

A quantitative analysis facilitates comparison of the amount of information required in the one-step approach (Figure 3.1, blue and brown bars) with the amount required in the two-step process presented here. This implies calculating the amount of information needed to plot $\Phi$ and $\Psi$ on the Ramachandran map (see the blue bars in Figure 3.1, and Table 3.1).

The one- and two-step models can be represented with respect to prediction of the (native) angles $(\Phi, \Psi)_N$ as follows:

One step: $AA \rightarrow (\Phi, \Psi)_N$

Two steps: $AA \rightarrow (\Phi, \Psi)_e \rightarrow (\Phi, \Psi)_z = (\Phi, \Psi)_N$

In order to illustrate this process, the angles $(\Phi, \Psi)$, $(\Phi, \Psi)_e$, and $(\Phi, \Psi)_z$ were calculated for a randomly chosen protein (1ADE, a ligase or synthetase)[4] for the purpose of determining the quantity of information required in the one- and two-step models. First, the quantity of information involved in the one-step prediction process was assessed by taking the sequence of amino acids and their appropriate probabilities of occurrence. The angles $\Phi$ and $\Psi$ were calculated for the 1ADE structure, their probability of occurrence was determined (at 5° intervals), and the quantity of information necessary for prediction of these dihedral angles was derived. All of these values are plotted on the *X* axis in Figure 3.4(a). Then the two-step model was analyzed, taking into account the conditional probability of finding the correct $(\Phi, \Psi)_z$, where the probability $p_z$ for these angles was treated as a

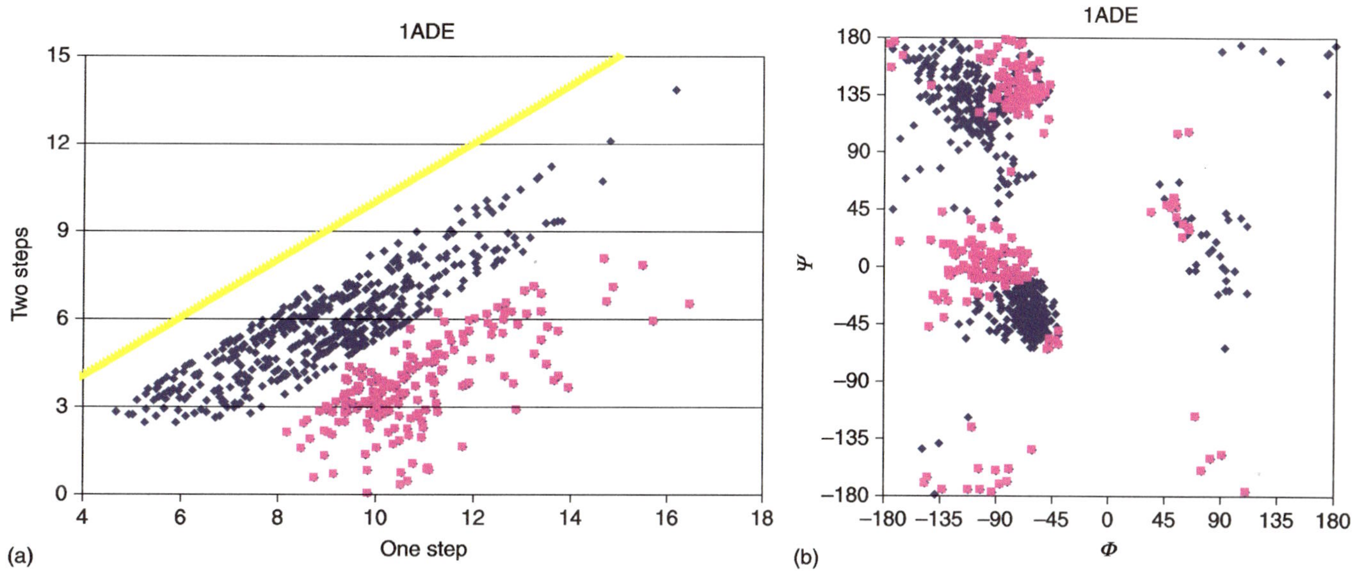


**Figure 3.4** Analysis of the structure of 1ADE, a randomly selected protein. (a) Quantity of information required in the one-step and two-step models. The diagonal $Y = X$ is also plotted to illustrate the relation between the two models. The pink points are points with significantly reduced information requirements. (b) Ramachandran map, with the angles $\Phi$ and $\Psi$ plotted for the protein 1ADE. Points for which the two-step model requires much less information than does the corresponding one-step model are marked in pink (compare the pink area in Figure 3.3(a))

conditional probability on the $(\Phi, \Psi)_e$ previously defined. The results are plotted on the *Y* axis in Figure 3.4(a).

As can be seen in Figure 3.4(a), the two-step model is consistently more efficient in terms of information than the one-step model. An interesting observation is that for certain points, the quantity of information is far below that for others. Further analysis indicated that these points represent angles $\Phi$ and $\Psi$ that belong mostly to zones D and F (see Figure 3.4(a) and (b)). In order to explain this phenomenon, the structural motifs corresponding to such pairs of angles were identified in the example protein 1ADE (L-lysine dehydrogenase).[5]

Figure 3.5 depicts this randomly selected protein 1ADE with fragments corresponding to zones D and F highlighted to showcase their structural specificity (note that these tend

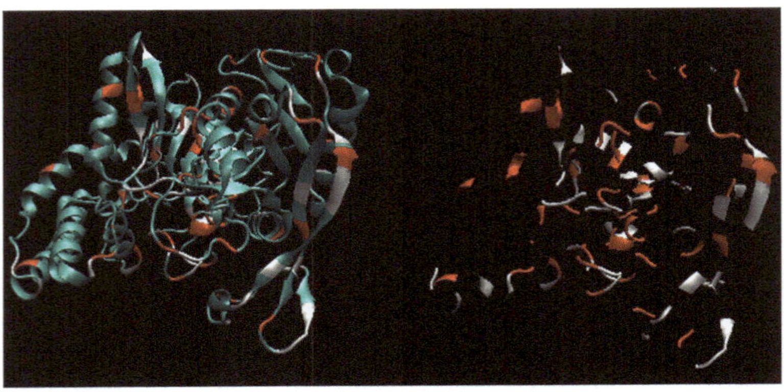

**Figure 3.5** 3D structural model of 1ADE. Fragments corresponding to zone F are marked in white, and fragments corresponding to zone D are marked in blue. Left: 3D representation of the entire protein; right: D (blue) and F (white) fragments extracted to highlight their structural properties. The program VMD was used to visualize the effects discussed



Published by Woodhead Publishing Limited, 2012

to correspond to terminal sections of well-ordered secondary structural motifs). A more detailed study, involving a larger group of proteins, suggests that these fragments can be described as unstructured loops. Codes D and F are typically represented by single, isolated amino acids (with the occasional pair of adjacent peptides), but it seems that such residues play a key part in determining the final shape of the protein. The fact that loop-like structures are particularly difficult to model correctly is a well-known problem in *in silico* folding simulations. By pinpointing amino acids that correspond to zones D and F at an early stage of the folding process, the accuracy of folding predictions can be significantly improved.

Zones A and B (and to some extent G) are poorly represented and, seemingly, highly specific. Their properties can be studied on the basis of the contingency table described in Chapter 1. Zones D and F are of special importance, as they often accompany well-defined secondary structures belonging to zones C and E, respectively. It is apparent, furthermore, that zones D and F are specifically associated with the initial and final sections of standard structural motifs. The role of zone F will be discussed in Chapter 8.

In summary, we can state that introducing the early-stage folding intermediate carries important benefits from the point of view of information theory. Determining the structure of this intermediate "consumes" the information present in the input sequence, enabling us to accurately model the resulting protein. Of particular note is the identification of zones D and F, which carry limited informational requirements but perform an important task by associating specific structural motifs with the termini of some basic structures found in the polypeptide chain.

Progressing beyond the early stage requires additional information, which can be obtained by studying the



Published by Woodhead Publishing Limited, 2012

environment in which the polypeptide chain undergoes folding (resulting in the formation of a hydrophobic core) and by taking into account the physiochemical properties of side chains responsible for stabilizing inter-residual interactions. In summary, we can posit that the amount of information carried by a sequence of amino acids is only sufficient to determine the structure of its ES intermediate and that additional information must come from the environment. Water – which is omnipresent as the inherent environment for all biological processes, including protein folding – is able to influence the folding process, resulting in the generation of a hydrophobic core. Thus, the missing information is "submitted" to the system by its aqueous environment. The possible role of other external factors is discussed in Chapter 4.

## 3.3 References


1. Shannon C.E. (1948) A mathematical theory of communication. *Bell Syst. Tech. J.* 27, 379–423, 623–56.
2. Jurkowski W., Wiśniowski Z., Konieczny L., Roterman I. (2004) The conformational sub-space in simulation of early-stage protein folding. *Proteins Struct. Funct. Bioinform.* 55, 115–27.
3. Brylinski M., Konieczny L., Roterman I. (2005) SPI – structure predictability index for protein sequences. *In Silico Biol.* 5(3), 227–37.
4. Silva M.M., Poland B.W., Hoffman C.R., Fromm H.J., Honzatko R.B. (1995) Refined crystal structures of unligated adenylosuccinate synthetase from *Escherichia coli*. *J. Mol. Biol.*, 254, 431–46.
5. Yoneda K., Fukuda J., Sakuraba H., Ohshima T. (2010) First crystal structure of L-lysine 6-dehydrogenase as an




Published by Woodhead Publishing Limited, 2012

NAD-dependent amine dehydrogenase. *J. Biol. Chem.* 285, 8444–53.

## 3.4 Suggested reading

Adler A., Youmaran R., Lionheart W.R. (2008) A measure of the information content of EIT data. *Physiol. Meas.* 29(6), S101–9.

Bacardit J., Stout M., Hirst J.D., Valencia A., Smith R.E., Krasnogor N. (2009) Automated alphabet reduction for protein datasets. *BMC Bioinform.* 10, 6.

Bouziane H., Messabih B., Chouarfia A. (2011) Profiles and majority voting-based ensemble method for protein secondary structure prediction. *Evol. Bioinform. Online* 7, 171–89.

Choi K., Kim S. (2011) Sequence-based enzyme catalytic domain prediction using clustering and aggregated mutual information content. *J. Bioinform. Comput. Biol.* 2011 9(5), 597–611.

Cong Q., Kinch L.N., Pei J., Shi S., Grishin V.N., et al. (2011) An automatic method for CASP9 free modeling structure prediction assessment. *Bioinformatics* 27(24), 3371–8.

De Baets G., Van Durme J., Reumers J., Maurer-Stroh S., Vanhee P., et al. (2012) SNPeffect 4.0: on-line prediction of molecular and structural effects of protein-coding variants. *Nucleic Acids Res.* 40(1), D935–9.

Desurvire E. (2000) Fundamental information-density limits in optically amplified transmission: an entropy analysis. *Opt. Lett.* 25(10), 701–3.

Fetrow J.S., Berg G. (1999) Using information theory to discover side chain rotamer classes: analysis of the effects of local backbone structure. *Pac. Symp. Biocomput.* 1999, 278–9.



Published by Woodhead Publishing Limited, 2012

Ghalwash M.F., Dunker A.K., Obradović Z. (2012) Uncertainty analysis in protein disorder prediction. *Mol. Biosyst.* 8(1), 381–91.

Gu F., Chen H., Ni J. (2008) Protein structural class prediction based on an improved statistical strategy. *BMC Bioinform.* 9 (Suppl. 6), S5.

Hamacher K. (2007) Information theoretical measures to analyze trajectories in rational molecular design. *J. Comput. Chem.* 28(16), 2576–80.

Horodecki M., Oppenheim J., Winter A. (2005) Partial quantum information. *Nature* 436(7051), 633–4.

King B.M., Tidor B. (2009) MIST: Maximum Information Spanning Trees for dimension reduction of biological data sets. *Bioinformatics* 25(9), 1165–72.

Li Y., Fang Y., Fang J. (2011) Predicting residue–residue contacts using random forest models. *Bioinformatics* 27(24), 3379–84.

Liu X., Krishnan A., Mondry A. (2005) An entropy-based gene selection method for cancer classification using microarray data. *BMC Bioinform.* 6, 76.

Martin L.C., Gloor G.B., Dunn S.D., Wahl L.M. (2005) Using information theory to search for co-evolving residues in proteins. *Bioinformatics* 21(22), 4116–24.

McGuffin L.J., Roche D.B. (2011) Automated tertiary structure prediction with accurate local model quality assessment using the IntFOLD-TS method. *Proteins* 79 (Suppl. 10), 137–46.

Mintseris J., Weng Z. (2004) Optimizing protein representations with information theory. *Genome Inform.* 15(1), 160–9.

Peng J., Xu J. (2011) RaptorX: exploiting structure information for protein alignment by statistical inference. *Proteins* 79 (Suppl. 10), 161–71.



Published by Woodhead Publishing Limited, 2012

Prado Martín F.M. del, Kosti A., Baayen R.H. (2004) Putting the bits together: an information theoretical perspective on morphological processing. *Cognition* 94(1), 1–18.

Reddy D.A., Prasad B.V., Mitra C.K. (2006) Comparative analysis of core promoter region: information content from mono and dinucleotide substitution matrices. *Comput. Biol. Chem.* 30(1), 58–62.

Rylance G.J., Johnston R.L., Matsunaga Y., Li C.B., Baba A., Komatsuzaki T. (2006) Topographical complexity of multidimensional energy landscapes. *Proc. Natl. Acad. Sci. USA* 103(49), 18551–5.

Shoombuatong W., Traisathit P., Prasitwattanaseree S., Tayapiwatana C., Cutler R., Chaijaruwanich J. (2011) Prediction of the disulphide bonding state of cysteines in proteins using conditional random fields. *Int. J. Data Min. Bioinform.* 5(4), 449–64.

Shultzaberger R.K., Roberts L.R., Lyakhov I.G., Sidorov I.A., Stephen A.G., et al. (2007) Correlation between binding rate constants and individual information of *E. coli* Fis binding sites. *Nucleic Acids Res.* 35(16), 5275–83.

Solis A.D., Rackovsky S. (2000) Optimized representations and maximal information in proteins. *Proteins* 38(2), 149–64.

Solis A.D., Rackovsky S. (2002) Optimally informative backbone structural propensities in proteins. *Proteins* 48(3), 463–86.

Stamps A.E. 3rd (2007) Entropy and environmental mystery. *Percept. Mot. Skills* 104(3 Pt. 1), 691–701.

Steward R.E., Thornton J.M. (2002) Prediction of strand pairing in antiparallel and parallel beta-sheets using information theory. *Proteins* 48(2), 178–91.



Published by Woodhead Publishing Limited, 2012

Su S.C., Lin C.J., Ting C.K. (2011) An effective hybrid of hill climbing and genetic algorithm for 2D triangular protein structure prediction. *Proteome Sci.* 9 (Suppl. 1), S19.

Sullivan D.C., Aynechi T., Voelz V.A., Kuntz I.D. (2003) Information content of molecular structures. *Biophys J.* 85(1), 174–90.

Swanson R., Kagiampakis I., Tsai J.W. (2008) An information measure of the quality of protein secondary structure prediction. *J. Comput. Biol.* 15(1), 65–79.

Swanson R., Vannucci M., Tsai J.W. (2009) Information theory provides a comprehensive framework for the evaluation for protein structure predictions. *Proteins* 74(3), 701–11.

Tlidi M., Taki M., Kolokolnikov T. (2007) Introduction: dissipative localized structures in extended systems. *Chaos* 17(3), 037101.

Wagner A. (2007) From bit to it: how a complex metabolic network transforms information into living matter. *BMC Syst. Biol.* 1, 33. doi:10.1186/1752-0509-1-33.

Zhang P., Obradovic Z. (2011) Unsupervised integration of multiple protein disorder predictors: the method and evaluation on CASP7, CASP8 and CASP9 data. *Proteome Sci.* (Suppl. 1), S12.

Zhong S., Moix J.M., Quirk S., Hernandez R. (2006) Dihedral-angle information entropy as a gauge of secondary structure propensity. *Biophys. J.* 91(11), 4014–23.



Published by Woodhead Publishing Limited, 2012

# The divergence entropy characterizing the internal force field in proteins

*M. Banach, D. Marchewka, M. Piwowar, and I. Roterman-Konieczna*

**Abstract:** This chapter deals with the characterization of the distribution of nonbonding interactions (including hydrophobic interactions) in the protein body. We establish that in most proteins, the hydrophobic interactions can be modeled by a 3D Gaussian function (the "fuzzy oil drop" model), while the electrostatic interactions follow a random distribution. A method for quantitative measurement of the degree of accordance between the expected and observed distributions is introduced and applied. From the point of view of optimizing nonbonding interactions, it is important to note that the non-Gaussian distributions observed in actual proteins approach the idealized Gaussian model when the residues involved in specific activities (enzymatic activity, ligand binding, protein complexation, etc.) are disregarded. It is concluded that pairwise (atom–atom) optimization is sufficient for modeling van der Waals forces and electrostatic interactions, whereas the optimization of hydrophobic interactions must reflect the tertiary structure of the molecule as a whole. It is shown that mismatches in the location of residues responsible for



Published by Woodhead Publishing Limited, 2012

interaction with external molecules distort the protein's hydrophobicity field, which would otherwise agree with the "fuzzy oil drop" model. The biological function of many proteins (enzymatic activity) is strongly related to deformations in their hydrophobic cores, which can be expressed quantitatively by the degree of structural divergence from the idealized "fuzzy oil drop" model. The deformations identified seem to be specific to specific biological functions.

**Key words:** hydrophobicity, oil drop model, divergence entropy, internal force field, random distribution, local versus global organization.

## 4.1 Internal force field for nonbonding interactions

As discussed earlier, some proteins (e.g., antifreeze and downhill proteins) contain hydrophobic cores whose structure closely mimics a 3D Gaussian distribution; however, other proteins diverge from this theoretical model. The basic interactions that stabilize protein structures include electrostatic and van der Waals forces. We should therefore consider whether such interactions can be modeled as internal force fields and whether they exhibit well-defined structural ordering.

In order to determine the distribution of the nonhydrophobic (i.e., electrostatic and van der Waals) interactions, a subset of PDB proteins was chosen for analysis using an energy optimization algorithm, eliminating any local collisions. Nonhydrophobic forces were calculated for each atom separately and then aggregated on the level of amino acids (using effective atoms to represent all interactions that



Published by Woodhead Publishing Limited, 2012

involve a particular side chain). The program Gromacs was used to perform the energy optimization and crystal relaxation procedure. All energy minimization calculations were performed using the Gromacs software package v4.0.3 and the Gromos96 43a1 force field.[1–5] The grouping option was used to tag each residue as a separate "group," interacting with the rest of the protein body.

Calculation of *O/T* and *O/R* values provided a basis upon which to determine the distribution of internal force fields throughout the molecule. Table 4.1 illustrates the distribution of nonbonding interactions by listing *O/T* and *O/R* values for the hydrophobic, electrostatic, and van der Waals forces.

It should be noted that 1HZC does not conform to the model presented here for any type of interaction. 1HZC represents the cold shock protein Bc-Csp from the thermophile *Bacillus caldolyticus*. The corresponding protein from the mesophile *Bacillus subtilis* is Bs-CspB.[6] This protein differs significantly

**Table 4.1** ***O/T* and *O/R* values expressing the accordance (boldface) or lack thereof (regular font) regarding hydrophobic, electrostatic, and van der Waals interactions in selected proteins**

| PDB ID | Hydrophobic | | Electrostatic | | Van der Waals | |
|---|---|---|---|---|---|---|
| | *O/T* | *O/R* | *O/T* | *O/R* | *O/T* | *O/R* |
| 1VII | **0.182** | **0.269** | 0.184 | 0.017 | 0.422 | 0.164 |
| 1BDC | **0.097** | **0.278** | 0.318 | 0.097 | **0.149** | **0.186** |
| 1RIJ | **0.110** | **0.231** | 0.244 | 0.110 | **0.080** | **0.087** |
| 2I5M | **0.139** | **0.253** | 0.444 | 0.215 | **0.116** | **0.188** |
| 1CSP | **0.114** | **0.208** | 0.266 | 0.036 | **0.103** | **0.190** |
| 1BDD | **0.136** | **0.276** | 0.388 | 0.118 | 0.216 | 0.204 |
| 1HZC | 0.213 | 0.207 | 0.323 | 0.057 | 0.308 | 0.252 |

from other antifreeze proteins[7] as it includes three specific mutations, namely R3E, E46A, and L66E, introduced in order to assess the impact of surface-bound amino acids on conformational stabilization. As it turns out, these mutations are sufficient to effect a loss of accordance between the observed hydrophobicity distribution and the "fuzzy oil drop" model (which correctly predicts the properties of other members of the antifreeze group).[7] In fact, the distribution observed in the mutated protein was close to random. This loss of accordance affects all types of interactions in this protein.

In all proteins belonging to the study group, the electrostatic interactions were distributed more or less randomly.

Good accordance between the observed and theoretical distributions of the van der Waals forces was observed in 1BDC,[8] 1RIJ,[9] 2I5M,[10] and 1CSP,[11] but was otherwise uncommon. To date, research indicates that only these four proteins are representative of the specific type of van der Waals interactions described here. Structural NMR analysis suggests that in 1BDC, hydrophobic residues congregate near the center of the molecule, as predicted by our model.[8] 1BDD is a crystalline form of 1BDC in which the van der Waals interactions do not conform to the "fuzzy oil drop" model (although the model remains applicable to the hydrophobicity distribution in 1BDC).[8]

1CSP is a cold shock protein in which an antiparallel five-stranded $\beta$-barrel (with three consecutive $\beta$-strands) creates a surface rich in aromatic and basic residues, which are presumably involved in bonding nucleic acids.[11] This protein is representative of a group of proteins in which both the hydrophobic and the van der Waals interactions remain well ordered.

1RIJ, a *de novo* protein, was synthesized while searching for a structure with perfect affinity for a subgroup consisting



Published by Woodhead Publishing Limited, 2012

of seven leucine residues.[9] According to experimental data, structural changes are triggered in this protein by complexation with the E6 viral protein. 1RIJ exhibits a highly ordered distribution of hydrophobic and van der Waals forces, consistent with the "fuzzy oil drop" model.

2I5M, a bacterial cold shock protein, is widely used as a baseline model in experimental and computational studies of protein stability.[10] Its high stability is evidenced in Table 4.1. The good agreement between the observed and theoretical distributions of the hydrophobic and van der Waals interactions (together with a random distribution of electrostatic interactions) yields further evidence in support of our model.

A random distribution of electrostatic interactions validates the pairwise (atom–atom) approach to optimization: the electrostatic energy remains constant regardless of the placement of each interacting pair of atoms within the protein body. On the other hand, applying the same rule to the van der Waals and hydrophobic interactions produces poor results (at least within the study group presented in this chapter). The increased strength of the van der Waals forces in the core is not simply a result of a greater atom density (otherwise a similar increase would be observed for the electrostatic interactions, which is not the case).

## 4.2 The impact of ligands

According to the assumption stated at the end of Chapter 3, the residues engaged in external interactions (binding ligands, forming complexes, etc.) need to be identified to allow the measurement of their influence on the hydrophobic core. The criterion for the identification of engagement of residues in complexation is the distance between the interacting



Published by Woodhead Publishing Limited, 2012

**Table 4.2** Hydrophobicity distributions in complete proteins, proteins with ligand-binding residues omitted ("no ligand"), proteins with protein–protein complexation residues omitted ("no P–P"), and proteins with ion- or DNA/RNA-binding residues omitted ("no ion, no RNA/DNA"). The *O/T* and *O/R* values in boldface indicate accordance with the model presented here

| PDB ID | Hydrophobic, complete | | Hydrophobic, no ligand | | Hydrophobic, no P–P | | Hydrophobic, no ion, no DNA/RNA | |
|---|---|---|---|---|---|---|---|---|
| | ***O/T*** | ***O/R*** | ***O/T*** | ***O/R*** | ***O/T*** | ***O/R*** | ***O/T*** | ***O/R*** |
| 1G57 | 0.171 | 0.160 | – | – | 0.171 | 0.159 | – | – |
| 1SQI | 0.197 | 0.191 | 0.191 | 0.190 | – | – | – | – |
| 1TK3 | 0.288 | 0.150 | 0.289 | 0.147 | – | – | – | – |
| 2Q20 | 0.143 | 0.133 | 0.147 | 0.138 | – | – | – | – |
| 1XJJ | 0.197 | 0.188 | **0.184** | **0.194** | – | – | – | – |
| 1AA4 | 0.218 | 0.205 | **0.207** | **0.213** | – | – | – | – |
| 3DRC | **0.154** | **0.184** | **0.149** | **0.193** | **0.155** | **0.185** | – | – |
| 1KHH | 0.147 | 0.144 | **0.140** | **0.150** | 0.138 | 0.129 | 0.176 | 0.165 |
| 1S99 | 0.172 | 0.170 | 0.176 | 0.174 | **0.163** | **0.180** | 0.173 | 0.172 |

molecules. In order to conform to PDBsum standards (http://www.ebi.ac.uk/pdbsum/), the threshold distance (indicating interaction) was assumed to be 2.9–3.9 Å.

Table 4.2 presents *O/T* and *O/R* values for dimeric forms and protein–ligand complexes. Note that in all such cases the force distributions remain close to random and do not exhibit Gaussian-like properties, except for 3DRC.

*O/T* and *O/R* values were calculated for generic molecules and for structures in which the residues responsible for complexation and ligand binding were excluded from the calculations. The purpose of this approach was to determine the causes of the observed deviations from the idealized core structure. It seems that in the case of 1S99,[7] the poor agreement between the theoretical and observed values is due to complexation with other protein molecules, while for 1KHH,[12] 1AA4,[13] and 1XJJ[14] it results from protein–ligand interactions.

## 4.3 Structures of homodimers – protein–protein interaction

Comparing the values listed in Tables 4.3 and 4.4 suggests two protein complexation mechanisms. The homodimers presented in Table 4.3 exhibit structural characteristics in accordance with the model presented here (involving a hydrophobic core) for each monomeric unit, as well as for the dimers. This can be interpreted as "static aggregation" of "fuzzy oil drops" in an aqueous environment, with little impact on the structure of each unit.

In contrast, Table 4.4 indicates that in the case of the proteins listed, reaching a distribution of hydrophobicity that corresponds to the "fuzzy oil drop" model requires complexation and the formation of a homodimer. This phenomenon may be



Published by Woodhead Publishing Limited, 2012

**Table 4.3** *O/T* and *O/R* values expressing the accordance of the hydrophobicity distribution

| PDB ID | Monomer A | | A, no P–P | | Monomer B | | B, no P–P | | Dimer | |
|---|---|---|---|---|---|---|---|---|---|---|
| | *O/T* | *O/R* | *O/T* | *O/R* | *O/T* | *O/R* | *O/T* | *O/R* | *O/T* | *O/R* |
| 1B8Z | 0.182 | 0.272 | 0.227 | 0.350 | 0.164 | 0.277 | 0.204 | 0.359 | 0.147 | 0.271 |
| 1BFM | 0.218 | 0.276 | 0.190 | 0.331 | 0.204 | 0.291 | 0.176M | 0.306 | 0.199 | 0.320 |
| 1C9O | 0.087 | 0.199 | 0.090 | 0.208 | 0.089 | 0.196 | 0.092 | 0.207 | 0.194 | 0.197 |
| 1F37 | 0.111 | 0.200 | 0.118 | 0.218 | 0.127 | 0.210 | 0.136 | 0.230 | 0.137 | 0.203 |
| 1RB7 | 0.122 | 0.226 | 0.121 | 0.231 | 0.122 | 0.226 | 0.118 | 0.225 | 0.205 | 0.217 |
| 1YE5 | 0.148 | 0.187 | 0.128 | 0.193 | 0.161 | 0.202 | 0.149 | 0.218 | 0.157 | 0.194 |
| 2Q3A | 0.116 | 0.183 | 0.105 | 0.191 | 0.137 | 0.173 | 0.122 | 0.186 | 0.165 | 0.172 |
| 2QD0 | 0.117 | 0.262 | 0.135 | 0.263 | 0.115 | 0.291 | 0.135 | 0.294 | 0.100 | 0.267 |
| 2VOC | 0.131 | 0.203 | 0.116 | 0.208 | 0.143 | 0.189 | 0.095 | 0.193 | 0.174 | 0.201 |
| 2WCI | 0.184 | 0.211 | 0.176 | 0.216 | 0.173 | 0.219 | 0.173 | 0.224 | 0.191 | 0.214 |

**Table 4.4** Examples of homodimers in which only the dimeric forms are in accordance with the model presented here (the *O/T* and *O/R* values for these are in boldface). The corresponding monomers do not reflect the idealized "fuzzy oil drop" structure, even when residues involved in complexation are eliminated

| PDB ID | Monomer A | | A, no P–P | | Monomer B | | B, no P–P | | Dimer | |
|---|---|---|---|---|---|---|---|---|---|---|
| | *O/T* | *O/R* | *O/T* | *O/R* | *O/T* | *O/R* | *O/T* | *O/R* | *O/T* | *O/R* |
| 1CMB | 0.333 | 0.201 | 0.326 | 0.225 | 0.283 | 0.208 | 0.286 | 0.239 | **0.212** | **0.214** |
| 1F4N | 0.238 | 0.171 | 0.274 | 0.157 | 0.252 | 0.171 | 0.322 | 0.180 | **0.179** | **0.188** |
| 1Y7Q | 0.314 | 0.254 | 0.302 | 0.289 | 0.298 | 0.282 | 0.292 | 0.318 | **0.230** | **0.317** |

termed "dynamic complexation." We can speculate that the original structure of such proteins follows the "fuzzy oil drop" model but that each monomer becomes distorted in the process of complexation, yielding finally a dimeric form that again contains a well-structured hydrophobic core, in accordance with the "fuzzy oil drop" model.

## 4.4 Protein containing a catalytic center

Similar calculations were performed for a protein containing catalytically active residues, namely 1J3N (EC 2.3.1.41, synthase II, or beta-keto-ACP synthase); see Table 4.5. 1J3N is an interesting case study owing to its dimeric character (yielding homodimers), complexation of a ligand (citric acid), multidomain composition, and catalytic biological activity.[15] The influence of each of these factors can be measured using this single protein as an example. The complete (monomeric) form of 1J3N exhibits no agreement with the model for all types of interactions. Eliminating residues responsible for ligand binding does not change the overall picture. The domain composition of 1J3N is also noteworthy: the first domain (1–259) is in accordance with the model, but the second domain diverges from it. One can speculate that the first domain is responsible for structural stabilization, acting as an "integrated chaperone" and providing a backbone for the entire molecule. The *O*/*T* and *O*/*R* values calculated for a derivative in which the residues responsible for protein–protein complexation have been eliminated show good agreement with the model as far as hydrophobic interactions are concerned. A similar phenomenon occurs upon elimination of catalytic residues. The electrostatic and van der Waals interactions remain universally divergent from the model.



Published by Woodhead Publishing Limited, 2012

**Table 4.5** **Properties of single-chain 1J3N (homodimer) showing that when residues involved in protein–protein complexation and catalytic activity are eliminated from calculation, the molecule remains in accordance with the "fuzzy oil drop" model. The complete protein does not have a clearly defined hydrophobic core; however, one of its two domains remains in agreement with the model. No accordance was found for any structural form regarding the electrostatic and van der Waals interactions**

| Protein 1J3N | Hydrophobic | | Electrostatic | | Van der Waals | |
|---|---|---|---|---|---|---|
| | *O/T* | *O/R* | *O/T* | O/R | *O/T* | *O/R* |
| Complete monomer | 0.167 | 0.166 | 0.378 | 0.037 | 0.245 | 0.071 |
| No ligand | 0.165 | 0.165 | 0.372 | 0.037 | 0.241 | 0.071 |
| No P–P | **0.159** | **0.178** | 0.394 | 0.040 | 0.246 | 0.071 |
| No catalytic residues | **0.165** | **0.273** | 0.379 | 0.168 | 0.379 | 0.035 |
| *Domains* | | | | | | |
| 1–259 | **0.144** | **0.173** | 0.378 | 0.040 | 0.248 | 0.069 |
| 260–408 | 0.200 | 0.157 | 0.408 | 0.012 | 0.232 | 0.073 |

Results for 1J3N are depicted in Figures 4.1 and 4.2. The electrostatic and hydrophobic interaction profiles for the entire molecule, as well as for derivatives in which the residues responsible for binding, complexation, and catalysis have been eliminated, indicate the impact of such residues on the hydrophobic core structure in 1J3N.

## 4.5 The role of exons

In searching for factors that stabilize tertiary structures and determine the shape of the hydrophobic core, we have



Published by Woodhead Publishing Limited, 2012

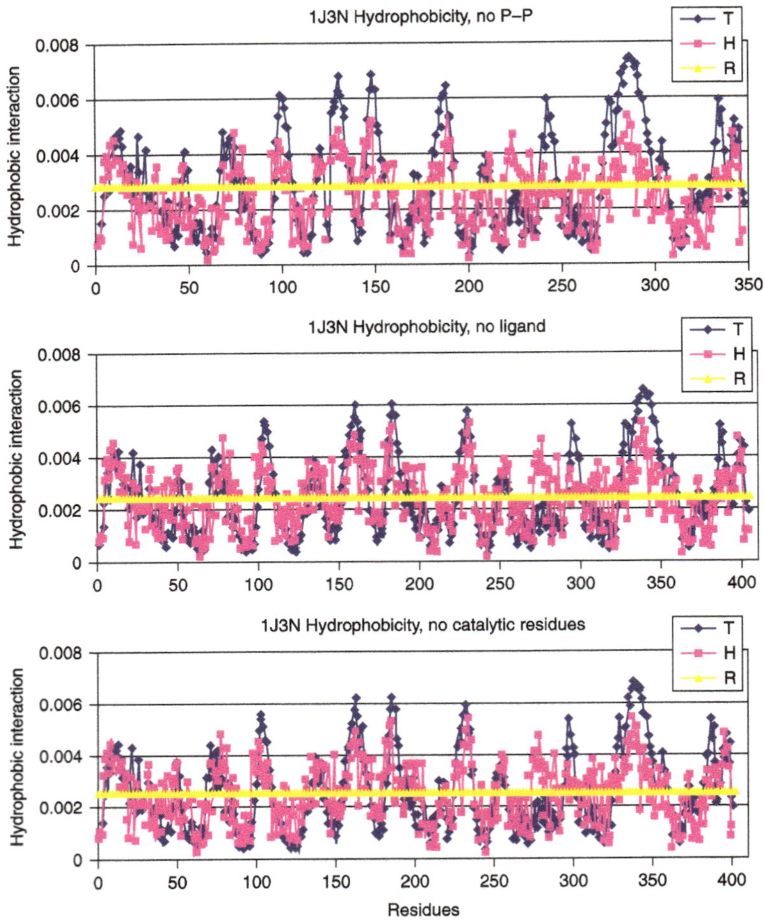


**Figure 4.1** Distribution of hydrophobic interactions in 1J3N without residues engaged in protein–protein interaction (top), without residues responsible for ligand complexation (center), and without catalytic residues (bottom). The top and bottom distributions appear to be in accordance with the model presented here (see Table 4.6). The dark blue rhombuses represent the theoretical distribution, the pink squares represent the observed distribution, and the yellow triangles indicate a random distribution



Published by Woodhead Publishing Limited, 2012

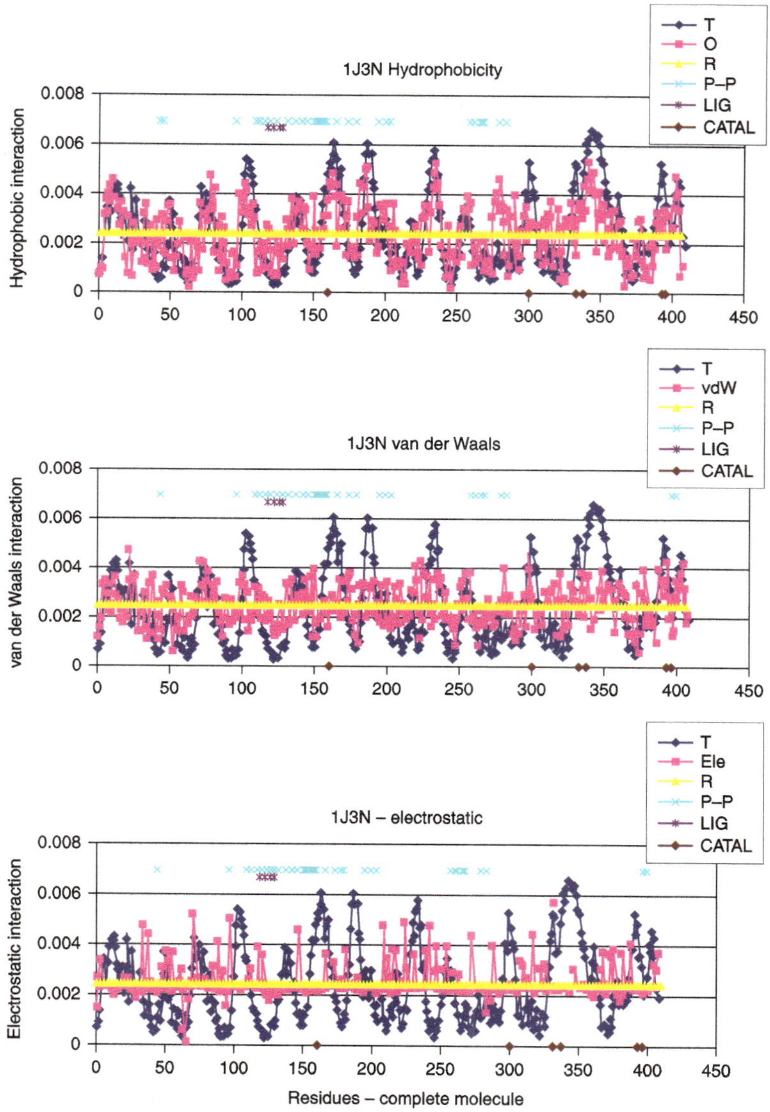


**Figure 4.2** **The influence of residues involved in protein–protein complexation on the distribution of hydrophobicity in 1J3N. The dark blue rhombuses represent the theoretical**

(*Continued*)



Published by Woodhead Publishing Limited, 2012

studied the distributions of hydrophobicity in exons in selected proteins. The results (obtained for a randomly selected group of proteins, both in accordance with and divergent from the model presented here) are presented in Table 4.6. The relatively high participation of exons in determining the shape of the hydrophobic core in structurally "accordant" proteins should come as no surprise; however, even in "discordant" proteins, there is typically at least one exon that significantly affects the core structure.

Analysis of the data presented in Table 4.6 suggests that low (or nonexistent) involvement of a given exon in binding ligands or in protein complexation is correlated with structural accordance of the encoded fragment of the polypeptide with the model presented here. Determining the role that each exon plays in stabilizing the protein's hydrophobic core and, consequently, its tertiary structure requires a large-scale case study, which is foreseen in the future.

**Figure 4.2** *caption continued*

**distribution, the pink squares represent the observed distribution, and the yellow triangles represent a random distribution. Residues involved in protein-protein complexation are tagged with light blue "X" signs, residues engaged in ligand binding are tagged with dark red "X" signs, and the brown rhombuses on the *X* axis correspond to catalytic residues. The red arrow in the top part of the figure indicates the boundary between the domains of 1J3N. The observed distributions of hydrophobic (top), van der Waals (center), and electrostatic (bottom) interactions are also shown (note the random character of the van der Waals and electrostatic distributions). The hydrophobic interactions in the first domain (1–259) appear to follow the theoretical model presented here (see Table 4.5)**

**Table 4.6** ***O/T*** **and** ***O/R*** **values for selected proteins and their exons. The values in bold indicate accordance with the predicted distribution. The second column lists the sequence numbers of the amino acids, along with the numbers of residues engaged in binding ligands (L), protein complexation (P), mutations (M), and enzymatic activity (E). The data in this column was derived from the PDBsum database**

| Protein | Exons | | Chain A | | Chain B | |
|---|---|---|---|---|---|---|
| | | | *O/T* | *O/R* | *O/T* | *O/R* |
| Proteins accordant with the model | | | | | | |
| ***2B8W*** | | | **0.194** | **0.198** | **0.189** | **0.196** |
| Exon 2 8 | 6–63 | L7 M1 P4 | **0.178** | **0.196** | **0.186** | **0.193** |
| Exon 3 12 | 65–106 | L7 M1 P8 | **0.157** | **0.185** | **0.104** | **0.188** |
| Exon 4 9 | 107–142 | P8 | **0.043** | **0.138** | **0.057** | **0.139** |
| Exon 5 12 | 144–195 | L3 P7 | 0.174 | 0.164 | 0.190 | 0.149 |
| Exon 6 12 | 197–276 | L3 P12 | 0.278 | 0.257 | **0.246** | **0.256** |
| Exon 7 0 | 278–293 | | **0.045** | **0.089** | 0.100 | 0.086 |
| ***1HTI*** | | | **0.175** | **0.187** | **0.182** | **0.194** |
| Exon 1 | 1–38 | P7 | **0.079** | **0.267** | 0.363 | 0.241 |
| Exon 2 | 40–79 | P18 | **0.165** | **0.183** | **0.152** | **0.183** |
| Exon 3 | 81–108 | P8 | 0.201 | 0.151 | **0.157** | **0.169** |
| Exon 4 | 109–152 | | **0.133** | **0.197** | **0.133** | **0.210** |
| Exon 5 | 154–181 | | 0.183 | 0.153 | **0.099** | **0.157** |
| Exon 6 | 182–210 | | 0.118 | 0.113 | 0.276 | 0.180 |
| Exon 7 | 212–248 | | **0.105** | **0.172** | **0.104** | **0.197** |
| ***2Z5E*** | | | **0.166** | **0.184** | **0.155** | **0.197** |
| Exon 1 | 1–72 | P23 | 0.197 | 0.195 | **0.175** | **0.211** |
| Exon 2 | 73–122 | | **0.091** | **0.168** | **0.086** | **0.175** |
| ***2J6W*** | | | **0.145** | **0.159** | 0.167 | 0.158 |
| Exon 4 | 50–89 | | 0.218 | 0.109 | 0.172 | 0.106 |
| Exon 5 | 90–141 | P2 | **0.111** | **0.209** | **0.152** | **0.194** |

(*Continued*)



Published by Woodhead Publishing Limited, 2012

**Table 4.6** (*Continued*)

| Protein | Exons | | Chain A | | Chain B | |
|---|---|---|---|---|---|---|
| | | | *O/T* | *O/R* | *O/T* | *O/R* |
| Exon 6 | 142–173 | P2 | **0.118** | **0.124** | 0.175 | 0.124 |
| Proteins discordant with the model | | | | | | |
| ***1XW5*** | | | 0.210 | 0.188 | 0.208 | 0.187 |
| Exon 1 | 1–11 | L2 E1 | 0.129 | 0.082 | 0.119 | 0.082 |
| Exon 2 | 12–36 | L1 | 0.199 | 0.203 | 0.203 | 0.200 |
| Exon 3 | 37–58 | L4 P2 | 0.182 | 0.152 | 0.204 | 0.150 |
| Exon 4 | 59–86 | L4 P10 | 0.173 | 0.133 | 0.172 | 0.130 |
| Exon 5 | 87–120 | L1 P9 | 0.276 | 0.243 | 0.262 | 0.245 |
| Exon 6 | 121–151 | P3 | **0.085** | **0.208** | **0.079** | **0.214** |
| Exon 7 | 152–188 | | 0.187 | 0.134 | 0.190 | 0.133 |
| Exon 8 | 189–217 | | **0.142** | **0.187** | **0.137** | **0.182** |
| ***1TK3*** | | | 0.289 | 0.150 | 0.304 | 0.150 |
| Exon 1 | – | | | | | |
| Exon 2 | – | | | | | |
| Exon 3 | 39–64 | | 0.256 | 0.143 | 0.266 | 0.162 |
| Exon 4 | 66–95 | L7 | 0.316 | 0.155 | 0.307 | 0.157 |
| Exon 5 | 96–122 | | 0.365 | 0.150 | 0.393 | 0.153 |
| Exon 6 | 123–139 | | 0.674 | 0.150 | 0.712 | 0.150 |
| Exon 7 | 141–164 | L4 | 0.244 | 0.135 | 0.252 | 0.150 |
| Exon 8 | 165–204 | L2 | 0.219 | 0.120 | 0.237 | 0.118 |
| Exon 9 | 206–258 | L6 P19 | 0.181 | 0.140 | 0.188 | 0.142 |
| Exon 10 | 259–332 | L8 P2 | 0.217 | 0.104 | 0.238 | 0.103 |
| Exon 11 | 334–423 | L2 | 0.296 | 0.152 | 0.301 | 0.146 |
| Exon 12 | 424–453 | | **0.153** | **0.167** | **0.144** | **0.158** |
| Exon 13 | 454–525 | L2 | 0.201 | 0.147 | 0.218 | 0.147 |
| Exon 14 | 526–569 | | 0.275 | 0.168 | 0.298 | 0.170 |
| Exon 15 | 571–604 | L2 | **0.128** | **0.131** | 0.139 | 0.132 |
| Exon 16 | 605–685 | E1 L1 P3 | 0.220 | 0.157 | 0.236 | 0.159 |
| Exon 17 | 687–716 | E2 P6 | 0.272 | 0.138 | 0.279 | 0.131 |



Published by Woodhead Publishing Limited, 2012

| | | | | | | |
|---|---|---|---|---|---|---|
| Exon 18 | 718–764 | E1 P19 | 0.435 | 0.116 | 0.459 | 0.108 |
| *1U3W* | | | 0.271 | 0.180 | 0.292 | 0.179 |
| Exon 1 | 1–5 | | **0.011** | **0.054** | **0.015** | **0.056** |
| Exon 2 | 6–39 | | **0.174** | **0.217** | **0.181** | **0.210** |
| Exon 3 | 40–86 | E3 L5 | 0.226 | 0.104 | 0.237 | 0.106 |
| Exon 4 | 87–114 | L2 P5 | 0.286 | 0.204 | 0.209 | 0.202 |
| Exon 5 | 115–188 | L4 | 0.177 | 0.145 | 0.203 | 0.148 |
| Exon 6 | 189–276 | L14 P5 | 0.228 | 0.210 | 0.244 | 0.207 |
| Exon 7 | 277–322 | L7 P23 | 0.527 | 0.156 | 0.510 | 0.157 |
| Exon 8 | 323–367 | L1 | **0.141** | **0.173** | **0.173** | **0.176** |
| Exon 9 | 368–376 | L1 | 0.110 | 0.044 | 0.112 | 0.047 |

## 4.6 Conclusions

Our analysis of various proteins with diverse characteristics of the hydrophobic core suggests the following conclusions:

1. The existence of proteins in which the hydrophobic core corresponds to the "fuzzy oil drop" model (downhill and antifreeze proteins)[2] proves that the aqueous environment has an influence on protein folding and reflects a situation in which certain residues are repelled by water. This mechanism can be exploited in *in silico* simulations of protein folding.
2. Applying the "fuzzy oil drop" model to folding simulations relies on pairwise (atom–atom) optimization of electrostatic interactions. This optimization needs to be performed for every pair of atoms, regardless of their location in the protein body.
3. Hydrophobic interactions should not be included in the optimization of the internal force field – rather, they should be subjected to a different optimization



Published by Woodhead Publishing Limited, 2012

algorithm, which minimizes the difference $\Delta\tilde{H}$ as defined in Eq. 2.6.

4. The van der Waals and electrostatic interactions should be optimized locally (most proteins exhibit poor agreement with the "fuzzy oil drop" model with respect to van der Waals interactions).
5. Describing the folding process with the use of a 3D Gaussian function appears justified, as the emerging hydrophobic core usually determines the protein's tertiary structure.
6. There are distortions in the structure of the hydrophobic core resulting from interactions with external molecules (ligand-binding or protein complexation processes).
7. The local deformations of the "fuzzy oil drop" caused by ligands may be interpreted as the result of adaptation to selected molecules, ensuring high specificity of the protein–ligand interactions.
8. The deformations of the hydrophobic core caused by catalytic residues (in enzymes) seem to correspond to the quantity of information required to accurately model the corresponding active sites. If the interaction with external fields is characterized by a low entropy (i.e., quantity of information), then the discrepancies between the idealized and observed hydrophobic-core structures may provide a basis upon which to model the enzyme's active sites.
9. Determining the cause, type, and extent of the divergence between the idealized and observed structures of the hydrophobic core remains an open issue, although the examples presented in this work provide hints as to the nature of the phenomena that occur.

Sample folding simulations taking account of the presence of an external field (described by a 3D Gaussian function that



Published by Woodhead Publishing Limited, 2012

triggers the formation of a hydrophobic core) have been conducted for BPTI,[16] lysozyme,[17] and T0215.[18] Similar computations involving the presence of a ligand have been performed for human hemoglobin[19] and ribonuclease.[20] The ligand was intended to affect the folding process in such a way as to ensure the creation of a suitable binding pocket. The effect of the external environment (the cellular membrane in this case) upon the dynamic properties of transmembrane proteins has been verified with the use of the "fuzzy oil drop" model.[21] The impact of mutations on the core structure (ranging from negligible to significant) has been studied using the example of antifreeze proteins.[7] Finally, the impact of mutations on the solvent-accessible area has been analyzed using support vector machines.[22]

Application of the "fuzzy oil drop" model to the identification of ligand binding sites[23] and active sites in hydrolases[24] has provided useful information concerning the properties of these proteins. Deformations in the structure of the hydrophobic core (resulting from interactions with external molecules) could be used to identify areas associated with biological activity. This issue was discussed further in Ref. 24, where a comparative study of various software packages used for identification of ligand-binding cavities was presented, describing their efficiency and accuracy. A similar comparative analysis of several packages (WebFEATURE,[25] SuMo,[26] ConSurf,[27] CASTp,[28,29] PASS,[30] and QSiteFinder)[31] using methods based on the "fuzzy oil drop" model can be found in Ref. 24.

We foresee a large-scale experiment in which an extensive set of proteins will be analyzed in order to determine the prevalence of the phenomena described in this chapter.



Published by Woodhead Publishing Limited, 2012

## 4.7 References

1. Berendsen H.J., van der Spoel D., van Drunen R. (1995) GROMACS: a message-passing parallel molecular dynamics implementation. *Comput. Phys. Commun.* 91, 43–56.
2. Berendsen H.J.C., Postma J.P.M., van Gunsteren W.F., Hermans J. (1981) Interaction models for water in relation to protein hydration. In Pullman B. (ed.), *Intermolecular Forces*, pp. 331–42. Dordrecht: Reidel.
3. Lindahl E., Hess B., van der Spoel D. (2001) GROMACS 3.0: a package for molecular simulation and trajectory analysis. *J. Mol. Model.* 7, 306–17.
4. van der Spoel D., Lindahl E., Hess B., Groenhof G., Mark A.E., Berendsen H.J. (2005) GROMACS: fast, flexible, and free. *J. Comput. Chem.* 26, 1701–18.
5. van der Spoel D., Lindahl E., Hess B., van Buuren A.R., Apol E., et al. Gromacs user manual, version 3.3. Available from: *http://www.gromacs.org* [accessed 7 May 2012].
6. Delbruck H., Mueller U., Perl D., Schmid F.X., Heinemann U. (2001) Crystal structures of mutant forms of the *Bacillus caldolyticus* cold shock protein differing in thermal stability. *J. Mol. Biol.* 313, 359–69.
7. Banach M., Prymula K., Jurkowski W., Konieczny L., Roterman I. (2012) Fuzzy oil drop model to interpret the structure of antifreeze proteins and their mutants. *J. Mol. Mod.* 18(1), 229–37.
8. Gouda H., Torigoe H., Saito A., Sato M., Arata Y., et al. (1992) Three-dimensional solution structure of the B domain of staphylococcal protein A: comparisons of the solution and crystal structures. *Biochemistry* 31, 9665–72.
9. Liu Y., Liu Z., Androphy E., Chen J., Baleja J.D. (2004) Design and characterization of helical peptides that



Published by Woodhead Publishing Limited, 2012

inhibit the E6 protein of papillomavirus. *Biochemistry* 43, 7421–31.

10. Max K.E., Wunderlich M., Roske Y., Schmid F.X., Heinemann U. (2007) Optimization of the gbeta1 domain by computational design and by in vitro evolution: structural and energetic basis of stabilization. *J. Mol. Biol.* 369, 1087–97.
11. Schindelin H., Marahiel M.A., Heinemann U. (1993) Universal nucleic acid-binding domain revealed by crystal structure of the *B. subtilis* major cold-shock protein. *Nature* 364, 164–8.
12. Komoto J., Huang Y., Takata Y., Yamada T., Konishi K., et al. (2002) Crystal structure of guanidinoacetate methyltransferase from rat liver: a model structure of protein arginine methyltransferase. *J. Mol. Biol.* 320, 223–35.
13. Fitzgerald M.M., Musah R.A., McRee D.E., Goodin D.B. (1996) A ligand-gated, hinged loop rearrangement opens a channel to a buried artificial protein cavity. *Nat. Struct. Biol.* 3, 626–31.
14. Larsson K.-M., Jordan A., Eliasson R., Reichard P., Logan D.T., Nordlund P. (2005) Structural mechanism of allosteric substrate specificity regulation in a ribonucleotide reductase. *Nat. Struct. Mol. Biol.* 11, 1142–9.
15. Bagautdinov B., Ukita Y., Miyano M., Kunishima N. (2008) Structure of 3-oxoacyl-(acyl-carrier protein) synthase II from *Thermus thermophilus* HB8. *Acta Crystallogr. Sect. F* 64, 358–66.
16. Bryliński M., Konieczny L., Roterman I. (2006) Hydrophobic collapse in late-stage folding (in silico) of bovine pancreatic trypsin inhibitor. *Biochimie* 88, 1229–39.
17. Bryliński M., Konieczny L., Roterman I. (2006) Fuzzy-oil-drop hydrophobic force field – a model to represent



Published by Woodhead Publishing Limited, 2012

late-stage folding (in silico) of lysozyme. *J. Biomol. Struct. Dynam.* 23, 519–27.

18. Konieczny L., Bryliński M., Roterman I. (2006) Gauss-function-based model of hydrophobicity density in proteins. *In Silico Biol.* 6, 2.
19. Brylinski M., Konieczny L., Roterman I. (2007) Is the protein folding an aim-oriented process? Human haemoglobin as example. *Int. J. Bioinform. Res. Appl.* 3, 234–60.
20. Bryliński M., Konieczny L., Roterman I. (2006) Hydrophobic collapse in (*in silico*) protein folding. *Comput. Biol. Chem.* 30, 255–67.
21. Zobnina V., Roterman I. (2009) Application of the fuzzy-oil-drop model to membrane protein simulation. *Proteins* 77, 378–94.
22. Dubey A., Realff M.J., Lee J.H., Bommarius A.S. (2006) Identifying the interacting positions of a protein using Boolean learning and support vector machines *Comput. Biol. Chem.* 30, 268–79.
23. Bryliński M., Prymula K., Jurkowski W., Kochańczyk M., Stawowczyk E., et al. (2007) Prediction of functional sites based on the fuzzy oil drop model. *PLoS Comput. Biol.* e94.
24. Prymula K., Jadczyk T., Roterman I. (2011) Catalytic residues in hydrolases: analysis of methods designed for ligand-binding site prediction. *J. Comput. Aided Mol. Des.* 25, 117–33.
25. Liang M.P., Banatao D.R., Klein T.E., Brutlag D.L., Altman R.B. (2003) WebFEATURE: an interactive web tool for identifying and visualizing functional sites on macromolecular structures. *Nucleic Acids Res.* 31, 3324–7.
26. Jambon M., Imberty A., Deléage G., Geourjon C.A. (2003) A new bioinformatic approach to detect



Published by Woodhead Publishing Limited, 2012

common 3D sites in protein structures. *Proteins* 52, 137–45.

27. Landau M., Mayrose I., Rosenberg Y., Glaser F., Martz E., et al. (2005) ConSurf 2005: the projection of evolutionary conservation scores of residues on protein structures. *Nucleic Acids Res.* 33, W299–302.
28. Binkowski A., Naghibzadeh S., Liang J. (2003) Castr: computed atlas of surface topography of proteins. *Nucleic Acids Res.* 31, 3352–5.
29. Dundas J., Ouyang Z., Tseng J., Binkowski A., Turpaz Y., et al. (2006) CASTp: computed atlas of surface topography of proteins with structural and topographical mapping of functionally annotated residues. *Nucleic Acids Res.* 34, W116–18.
30. Brady G.P., Stouten P.F. (2000) Fast prediction and visualization of protein binding pockets with PASS. *J. Comput. Aided Mol. Des.* 14, 383–401.
31. Laurie A.T., Jackson R.M. (2005) QSiteFinder: an energy-based method for the prediction of protein–ligand binding sites. *Bioinformatics* 21: 1908–16.



Published by Woodhead Publishing Limited, 2012

5

# Ligand-binding-site recognition

*M. Banach, L. Konieczny, and I. Roterman-Konieczna*

**Abstract:** Analysis of ligand complexation sites according to the "fuzzy oil drop" model implies a search for specific distortions in the protein's internal hydrophobicity distribution field. A quantitative measure of these distortions enables analysts to determine the protein's capacity for binding ligands. However, this relation is not always clear and evident in all protein–ligand complexes. While the existence of proteins in accordance with the model supports our theoretical assumptions, we must conclude that there is, in fact, more than one mechanism by which proteins are capable of binding ligands. It appears that the list of external factors that affect the distribution of internal hydrophobicity is not strictly limited to potential ligands (i.e., small molecules) but includes objects such as membranes, metal ions, and cellular organelles whose proximity deforms the protein's hydrophobic core in specific ways.

**Key words:** ligand complexation, binding pocket, divergence entropy, ROC curve.



Published by Woodhead Publishing Limited, 2012

## 5.1 General model

In the previous chapter, it was assumed that the structure of each binding site can be mapped to a specific deformation of the core, as reflected by the protein's profile. This suggests that it may be possible to identify ligand binding sites through an analysis of discrepancies between the observed and theoretical distributions of hydrophobicity in the protein's core.

In order to enable such an analysis, we will present examples of proteins in which ligand binding sites can be accurately identified, as well as proteins that do not conform to the model presented here. As mentioned before, it is assumed that local maxima of the $\Delta\tilde{H}$ profile represent hydrophobicity deficiencies (compared with the "idealized" distribution), whereas residues corresponding to local minima of the $\Delta\tilde{H}$ profile (i.e., having excess hydrophobicity) are suspected of involvement in ligand complexation. In order to validate our model, we have conducted analyses of proteins that form complexes with relatively large ligands ($NAD^+$ and FMN particles). The accuracy of ligand-binding-site predictions has been verified by the use of receiver operating characteristic (ROC) curves.

## 5.2 ROC curves

Receiver operating characteristic curves[1] were applied to measure the accordance of selected protein–ligand complexes with the assumed model. True and false positive rates were calculated for each curve using a naive algorithm, based on the $\Delta\tilde{H}_i$ values obtained and maps of actual contacts (which identified residues involved in ligand binding). The algorithm was applied independently to two residue sets, namely residues with positive and negative $\Delta\tilde{H}_i$ values, resulting in



Published by Woodhead Publishing Limited, 2012

two ROC curves for each chain. The algorithm operated as follows:

1. For each set, the algorithm iteratively adjusted the cutoff level for $\Delta\tilde{H}_i$ (starting from the largest absolute value and then approaching 0) and classified all residues with $\Delta\tilde{H}_i$ values above this level as involved in ligand complexation.
2. The true positive ratio (TPR) was calculated as the ratio between the number of residues correctly tagged in step 1 as involved in complexation and the total number of residues actually involved in ligand complexation, for a given residue set. This quantity was plotted on the *Y* axis.
3. The false positive ratio (FPR) was calculated as the ratio between the number of residues incorrectly tagged as involved in ligand complexation and the total number of residues not involved in such activity. This variable was plotted on the *X* axis.
4. The results from the previous steps of the algorithm were progressively aggregated to ensure a monotonic shape of the ROC curves.

The faster the TPR approaches 1 (along with a corresponding minimization of the false negative rate), the better the accordance of the assumed model with the experimental observations for a particular protein–ligand complex.

These ROC curves were characterized by two parameters that enabled automatic identification of proteins that were in accordance with the model presented here. We divided the area between the ROC curve and the diagonal $Y = X$ by the maximum possible value of this area (half of the unit square) to yield a percentage value; if this percentage was large, it indicated good agreement between the observations and theoretical predictions. Positive areas were associated with

ROC curves above the diagonal, and negative values with curves below the diagonal. Chains were interpreted as being in accordance with the model when any of their ROC curves filled at least 30% of the "positive" area. The second parameter was the percentage of residues involved in ligand complexation for a given ROC curve, compared with the total number of amino acids in the structure that exhibited this property.

Computations were performed separately for $\Delta\tilde{H}_i \leq 0$ and $\Delta\tilde{H}_i \geq 0$. The former case describes protein–ligand complexation mediated by residues with excess hydrophobicity, while the latter case corresponds to residues characterized by a hydrophobicity deficiency, which can be assumed to be ready for protein–protein complexation. Information about the actual involvement of specific residues in ligand complexation was taken from PDBsum, which serves as a "gold standard" in this field.[2]

One of the proteins in which FMN complexation sites can be accurately predicted is the chain 1VYR (pentaerythritol tetranitrate reductase complexed with the nitroaromatic substrate 2,4,6-trinitrophenol (picric acid)). This protein and its complex with the ligand are of special interest because the ligand is associated with enzymatic cavity.[3] Figure 5.1 illustrates the identification of the residues involved in binding the ligand.

In contrast, 3F2V (a protein with an unknown biological role, discovered in the Structural Genomics project)[4] does not permit easy recognition of FMN binding sites. In this case, ligand binding is mediated by residues with varying $\Delta\tilde{H}$ values, as depicted in Figure 5.2.

Correct identification of the $NAD^+$ binding pocket is possible in the protein 1LJ8, an oxidoreductase (mannitol dehydrogenase).[5] Results obtained by applying the "fuzzy oil drop" model are presented in Figure 5.3. It is worth noting



Published by Woodhead Publishing Limited, 2012

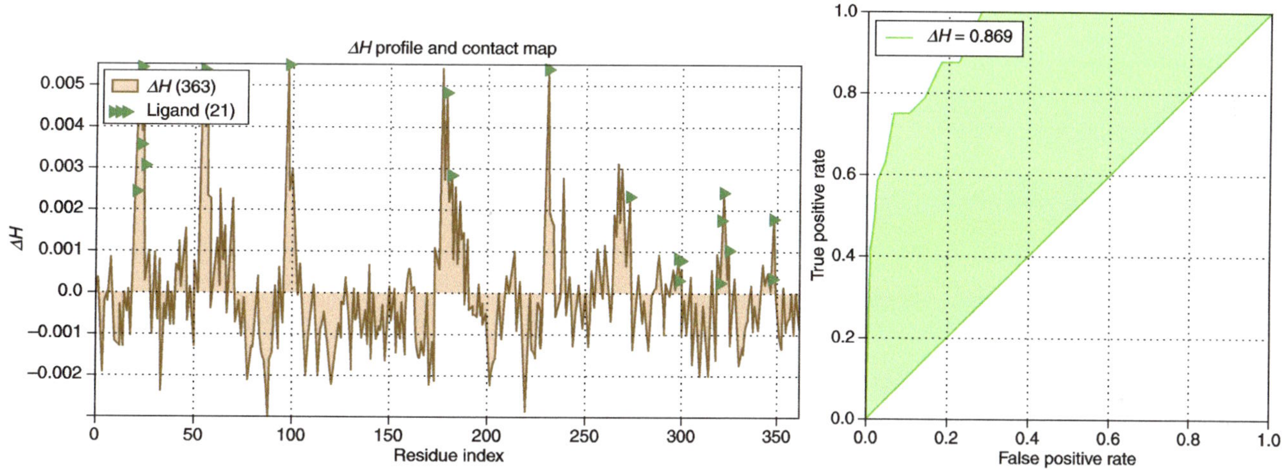


**Figure 5.1** $\Delta\tilde{H}$ profile of the protein 1VYR (left) and the corresponding ROC curve for FMN binding-site recognition (right). The involvement of residues associated with local maxima results in a rapid increase in the TPR, particularly for high $\Delta\tilde{H}$ values. The green triangles in the left-hand plot indicate residues engaged in ligand complexation

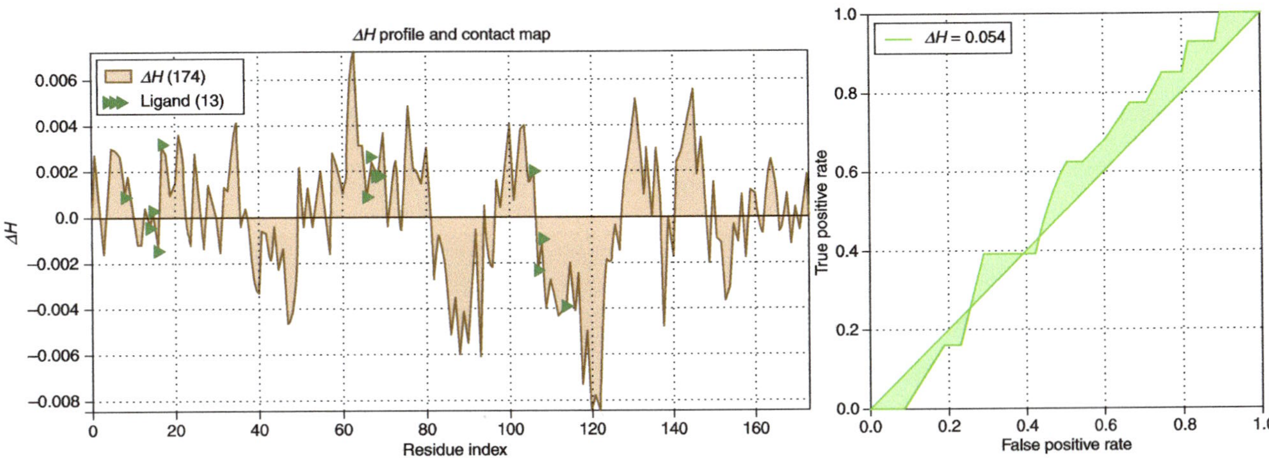


**Figure 5.2** $\Delta\tilde{H}$ profile of the protein 3F2V (left) and the corresponding ROC curve for FMN binding-site recognition (right). The involvement of residues with various $\Delta\tilde{H}$ values (including negative values) results in an ROC curve that closely approximates the diagonal. The $\Delta\tilde{H}$ values (left) represent the area between the diagonal and the curve

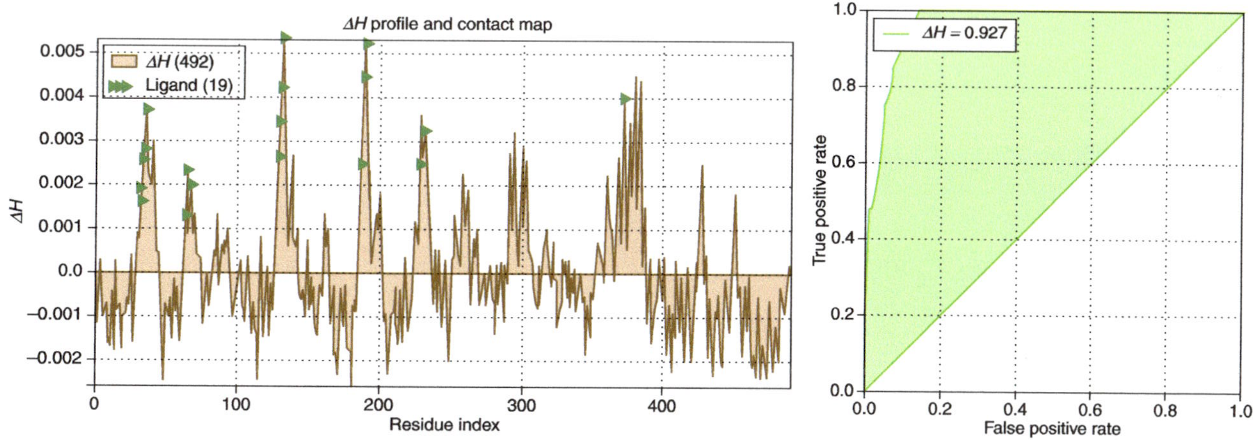


**Figure 5.3** $\Delta\tilde{H}$ profile of the protein 1LJ8 (left) and the corresponding ROC curve for $NAD^+$ ligand-binding-site recognition (right). The involvement of residues associated with local maxima results in a rapid increase in the TPR, particularly for high $\Delta\tilde{H}$ values

that a clear majority of the residues associated with local maxima in the profile are actually involved in binding ligands.

In order to highlight the successes and failures of the "fuzzy oil drop" model in predicting ligand binding sites, we now present the protein 1TOX as an example of a structure for which the model produces incorrect results. 1TOX is a *Diphtheria* toxin dimer complexed with $NAD^+$.[6] Its $\Delta\tilde{H}$ profile and the corresponding ROC curve are depicted in Figure 5.4. As can be seen, ligand binding is mediated by residues with both high and low $\Delta\tilde{H}$ values, which means that the $\Delta\tilde{H}$ profile is of limited use for binding-site recognition.

As mentioned earlier, we expect hydrophobic ligands to attach themselves to residues representing local hydrophobicity deficiencies (maxima in the $\Delta\tilde{H}$ profile) on the protein surface. However, it turns out that 1B1C (an oxidoreductase) instead binds ligands via residues with excess hydrophobicity (minima in the $\Delta\tilde{H}$)[7] (Figure 5.5).

In the case of 1VYR and 1LJ8, the local hydrophobicity deficiencies are compensated for by binding hydrophobic ligands. However, in 1B1C, the protein molecule isolates an area of excess hydrophobicity by attaching itself to the hydrophobic fragment of the ligand, thus shielding it from contact with water. It is also interesting to note that the ligand in question is the FMN molecule, which in most other cases interacts with residues associated with local maxima in the $\Delta\tilde{H}$ profile.

Table 5.1 compares *O*/*T* and *O*/*R* values for the proteins described above. In addition, the table lists *O*/*T* and *O*/*R* values for chains deprived of ligand-binding residues, and the area between the ROC curve and the corresponding diagonal on the plot of TPR versus FPR.

The proteins in which the hydrophobic core closely matches the "fuzzy oil drop" model can usually be modeled



Published by Woodhead Publishing Limited, 2012

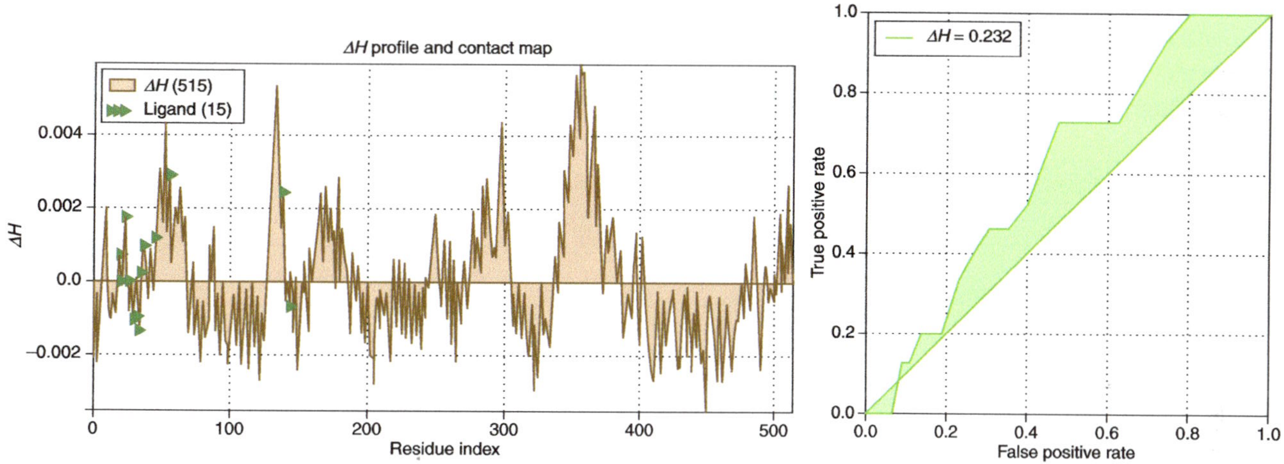


**Figure 5.4** $\Delta\tilde{H}$ profile of the protein 1TOX (left) and the corresponding ROC curve for $NAD^+$ ligand-binding-site recognition (right). The involvement of residues with various $\Delta\tilde{H}$ values (including negative values) results in an unfavorable ROC curve

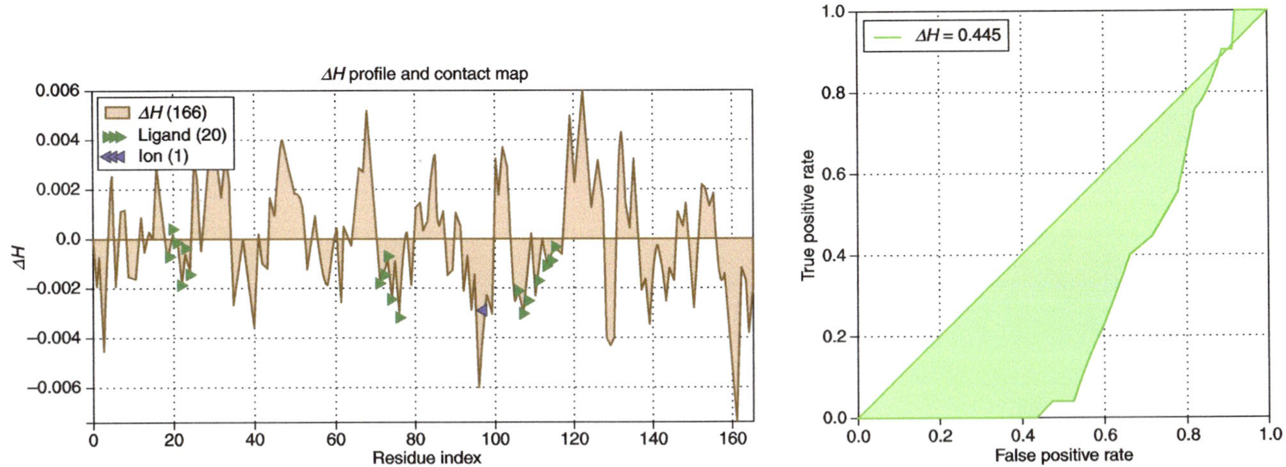


**Figure 5.5** $\Delta\tilde{H}$ profile of the protein 1B1C (left) and the corresponding ROC curve for ligand-binding-site recognition (right). The involvement of residues associated with local maxima results in a rapid increase in the TPR, particularly for high $\Delta\tilde{H}$ values. The ROC curve is plotted below the diagonal, representing negative $\Delta\tilde{H}$ values

**Table 5.1** ***O/T*** **and** ***O/R*** **values for complete protein chains and for parts of the chain not involved in binding ligands. Instances where** $O/T < O/R$ **are shown in bold. The rightmost column presents the results of the ROC analysis, i.e., the area between the diagonal and the curve as a percentage of the total area of the corresponding triangle on the ROC chart**

| PDB ID | Complete protein | | No ligand | | ROC curve (%) |
|---|---|---|---|---|---|
| | *O/T* | *O/R* | *O/T* | *O/R* | |
| 1VYR | **0.1521** | **0.1540** | **0.1205** | **0.1586** | **86.9** |
| 3F2V | 0.2180 | 0.1506 | 0.2290 | 0.1523 | 5.4 |
| 1LJ8 | 0.3183 | 0.1425 | 0.2891 | 0.1441 | **92.7** |
| 1TOX | 0.4848 | 0.1798 | 0.4891 | 0.1802 | 23.2 |
| 1B1C | **0.1226** | **0.2068** | **0.1307** | **0.2210** | **44.5** |

using ROC curves. An exception to this rule is 1LJ8, where correct results are produced by the ROC curve analysis despite the fact that this molecule does not conform to the "fuzzy oil drop" model. A possible explanation of this phenomenon is that the ligand molecule may actively participate in the protein folding process, ensuring the generation of a highly specific binding pocket. The presence of a binding pocket in 1LJ8 distorts its hydrophobic core to a significant degree. This distortion allows us to accurately identify the residues involved in ligand binding (as reflected by ROC curve analysis).

## 5.3 Summary

The assumption that irregularities in the structure of the protein's hydrophobic core can be triggered by the presence



Published by Woodhead Publishing Limited, 2012

of ligands has been verified by analyzing several different proteins. It is assumed that residues with local hydrophobicity deficiencies (local maxima in the $\Delta\tilde{H}$ profile) form the ligand binding site and that the ligand compensates for such deficiencies by protecting these residues from contact with water. Our database of homodimers contains examples of proteins that support this argument. However, it also includes homodimers that bind ligands by means of residues with excess hydrophobicity (see Chapter 6). A possible interpretation of this fact is that the ligand shields the protein's hydrophobic surface from the aqueous environment.

It should be noted that the analysis presented here can also be applied to the identification of enzymatic active sites, for instance in hydrolases. Detailed results can be found in reference.[8] Comparative analysis places our model as the second or third best (depending on the criteria used) among 10 programs used to identify active sites.[9–15] The examples discussed in this chapter support the model's validity, although they also suggest that ligand complexation mechanisms are highly diverse and dependent on the protein's specific biological role. Thus, we should not expect the model to be applicable to the vast majority of proteins.

## 5.4 References

1. Fawcett T. (2006) An introduction to ROC analysis. *Pattern Recognit. Lett.* 27, 861–4.
2. EMBL-EBI. PDBsum. Available from: *http://www.ebi.ac.uk/pdbsum/* [accessed 7 May 2012].
3. Khan H., Barna T., Harris R.J., Bruce N.C., Barsukov I., et al. (2004) Atomic resolution structures and solution behavior of enzyme–substrate complexes of *Enterobacter cloacae* PB2 pentaerythritol tetranitrate reductase.



Published by Woodhead Publishing Limited, 2012

Multiple conformational states and implications for the mechanism of nitroaromatic explosive degradation. *J. Biol. Chem.* 279, 30563–72.

4. Forouhar F., Lew S., Seetharaman J., Janjua H., Sahdev S., et al. Crystal structure of the general stress protein 14 (TDE0354) in complex with FMN from *Treponema denticola*, Northeast Structural Genomics Consortium Target TdR58. Available from: *http://www.rcsb.org/pdb* [accessed 7 May 2012].
5. Kavanagh K.L., Klimacek M., Nidetzky B., Wilson D.K. (2002) Crystal structure of *Pseudomonas fluorescens* mannitol 2-dehydrogenase binary and ternary complexes. Specificity and catalytic mechanism. *J. Biol. Chem.* 277, 43433–42.
6. Bell C.E., Eisenberg D. (1996) Crystal structure of *Diphtheria* toxin bound to nicotinamide adenine dinucleotide. *Biochemistry* 35, 1137–49.
7. Zhao Q., Modi S., Smith G., Paine M., McDonagh P.D., et al. (1999) Crystal structure of the FMN-binding domain of human cytochrome P450 reductase at 1.93 Å resolution. *Protein Sci.* 8, 298–306.
8. Prymula K., Jadczyk T., Roterman I. (2011) Catalytic residues in hydrolases: analysis of methods designed for ligand-binding site prediction. *J. Comput. Aided Mol. Des.* 25(2), 117–33.
9. Liang M.P., Banatao D.R., Klein T.E., Brutlag D.L., Altman R.B. (2003) WebFEATURE: an interactive web tool for identifying and visualizing functional sites on macromolecular structures. *Nucleic Acids Res.* 31, 3324–7.
10. Jambon M., Imberty A., Deléage G., Geourjon C.A. (2003) A new bioinformatic approach to detect common 3D sites in protein structures. *Proteins* 52, 137–45.
11. Landau M., Mayrose I., Rosenberg Y., Glaser F., Martz E., et al. (2005) ConSurf 2005: the projection of



Published by Woodhead Publishing Limited, 2012

evolutionary conservation scores of residues on protein structures. *Nucleic Acids Res.* 33, W299–302.

12. Binkowski A., Naghibzadeh S., Liang J. (2003) Castr: computed atlas of surface topography of proteins. *Nucleic Acids Res.* 31, 3352–5.
13. Dundas J., Ouyang Z., Tseng J., Binkowski A., Turpaz Y., et al. (2006) CASTp: computed atlas of surface topography of proteins with structural and topographical mapping of functionally annotated residues. *Nucleic Acids Res.* 34, W116–18.
14. Brady G.P., Stouten P.F. (2000) Fast prediction and visualization of protein binding pockets with PASS. *J. Comput. Aided Mol. Des.* 14, 383–401.
15. Laurie A.T., Jackson R.M. (2005) QSiteFinder: an energy-based method for the prediction of protein–ligand binding sites. *Bioinformatics* 21, 1908–16.

## 5.5 Suggested reading

Daggett V., Fersht A.R. (2009) Protein folding and binding: moving into unchartered territory. *Curr. Opin. Struct. Biol.* 19(1), 1–2.

Di Cera E., Bassi F.A., Gill S.J. (1989) Information theory and the analysis of ligand-binding data. *Biophys. Chem.* 34(1), 19–28.

Erill I., O'Neill M.C. (2009) A reexamination of information theory-based methods for DNA-binding site identification. *BMC Bioinform.* 10, 57.

Fornili A., Autore F., Chakroun N., Martinez P., Fraternali F. (2012) Protein–water interactions in MD simulations: POPS/POPSCOMP solvent accessibility analysis, solvation forces and hydration sites. *Methods Mol. Biol.* 819, 375–92.



Published by Woodhead Publishing Limited, 2012

Key J., Scheuermann T.H., Anderson P.C., Daggett V., Gardner K.H. (2009) Principles of ligand binding within a completely buried cavity in HIF2alpha PAS-B. *J. Am. Chem. Soc.* 131(48), 17647–54.

Kulharia M., Goody R.S., Jackson R.M. (2008) Information theory-based scoring function for the structure-based prediction of protein–ligand binding affinity. *J. Chem. Inf. Model.* 48(10), 1990–8.

Kuzu G., Keskin O., Gursoy A., Nussinov R. (2012) Expanding the conformational selection paradigm in protein–ligand docking. *Methods Mol. Biol.* 819, 59–74.

Rutherford K., Daggett V. (2009) The V119I polymorphism in protein *L*-isoaspartate *O*-methyltransferase alters the substrate-binding interface. *Protein Eng. Des. Sel.* 22(12), 713–21.

Salonen L.M., Holland M.C., Kaib P.S., Haap W., Benz J., et al. (2011) Molecular recognition at the active site of factor Xa: cation–$\pi$ interactions, stacking on planar peptide surfaces, and replacement of structural water. *Chemistry* 2012 Jan 2; 18(1): 213–22.



Published by Woodhead Publishing Limited, 2012

6

# Use of the "fuzzy oil drop" model to identify the complexation area in protein homodimers

*M. Banach, L. Konieczny, and I. Roterman-Konieczna*

**Abstract:** The aim of this chapter is to demonstrate how an analysis of the hydrophobicity distribution in homodimers can help identify complexation areas in each participating protein. Irregularities in the observed distribution of hydrophobicity, when compared with the expected, well-ordered distribution, may be treated as criteria for recognition of complexation sites. The reliability of this approach is verified using ROC diagrams. The relation between the true positive ratio and the false positive ratio calculated for local hydrophobicity maxima (deficiencies) and minima (excesses) shows that the model matches experimental observations. A case study involving a set of sample proteins is also presented.

**Key words:** hydrophobicity, protein–protein complexation, oil drop, fuzzy oil drop

## 6.1 General description

In order to validate our proposed method for the identification of complexation sites, a sample set of homodimers was prepared by scanning the PDB database for occurrences of the keyword "HOMODIMER." Structures that did not consist of exactly two chains or that occurred in complexes with DNA were excluded from the analysis. In addition, the Needleman–Wunsch[1] alignment algorithm was applied to verify sequence similarity (or identity): dimers with chains differing by more than 20 amino acids (by substitutions, insertions, or deletions) were discarded, resulting in a set of 208 acceptable homodimers. The selection was based upon the PDB as it existed in March 2010.[2]

The list of proteins selected for analysis is presented in Table 6.1. An in-depth explanation of the classification criteria will be given later in this chapter.

**Table 6.1** **Kullback–Leibler entropy calculated for the proteins (monomers and dimers) studied. The relation $O/T < O/R$ is treated as an indication of accordance. The values in bold indicate "accordant" cases**

| PDB ID | *O/T* (1) | *O/R* (1) | *O/T* (2) | *O/R* (2) | *O/T* (dimer) | *O/R* (dimer) |
|---|---|---|---|---|---|---|
| Both monomers accordant, dimer accordant | | | | | | |
| 1B8Z | **0.1819** | **0.2725** | **0.1644** | **0.2764** | **0.1474** | **0.2711** |
| 1BFM | **0.2183** | **0.2765** | **0.2045** | **0.2912** | **0.1987** | **0.3200** |
| 1C9O | **0.0870** | **0.1996** | **0.0892** | **0.1963** | **0.1943** | **0.1968** |
| 1F37 | **0.1110** | **0.2010** | **0.1275** | **0.2106** | **0.1376** | **0.2026** |
| 1RB7 | **0.1224** | **0.2261** | **0.1223** | **0.2226** | **0.2058** | **0.2178** |
| 1YE5 | **0.1480** | **0.1876** | **0.1608** | **0.2016** | **0.1573** | **0.1939** |
| 2Q3A | **0.1161** | **0.1832** | **0.1368** | **0.1733** | **0.1653** | **0.1718** |
| 2VOC | **0.1312** | **0.2033** | **0.1133** | **0.1891** | **0.1742** | **0.2008** |
| 2WCI | **0.1844** | **0.2106** | **0.1728** | **0.2191** | **0.1910** | **0.2143** |



Published by Woodhead Publishing Limited, 2012

| Both monomers accordant, dimer not accordant | | | | | | |
|---|---|---|---|---|---|---|
| 1AC6 | **0.0847** | **0.1744** | **0.0842** | **0.1792** | 0.2929 | 0.1749 |
| 1BEB | **0.1555** | **0.2400** | **0.1508** | **0.2295** | 0.2548 | 0.2331 |
| 1BFT | **0.1640** | **0.1972** | **0.1699** | **0.1992** | 0.2643 | 0.1948 |
| 1B04 | **0.1442** | **0.1887** | **0.1441** | **0.1927** | 0.3463 | 0.1779 |
| 1F05 | **0.1880** | **0.1911** | **0.1912** | **0.1938** | 0.4460 | 0.1882 |
| 1F9Z | **0.1756** | **0.1911** | **0.1691** | **0.1897** | 0.1971 | 0.1778 |
| 1GOH | **0.1753** | **0.2122** | **0.1984** | **0.2083** | 0.3103 | 0.2073 |
| 1HDF | **0.0739** | **0.1334** | **0.0717** | **0.1385** | 0.2273 | 0.1359 |
| 1HTI | **0.1749** | **0.1873** | **0.1823** | **0.1942** | 0.3885 | 0.1780 |
| 1JUB | **0.1683** | **0.1735** | **0.1688** | **0.1748** | 0.2381 | 0.1659 |
| 1K66 | **0.1684** | **0.1714** | **0.1701** | **0.1739** | 0.2601 | 0.1694 |
| 1K68 | **0.1574** | **0.1665** | **0.1517** | **0.1571** | 0.4144 | 0.1597 |
| 1MY7 | **0.1455** | **0.2019** | **0.1692** | **0.1889** | 0.2432 | 0.1972 |
| 1QLL | **0.1622** | **0.2426** | **0.1551** | **0.2436** | 0.5226 | 0.2397 |
| 1RRG | **0.1545** | **0.1834** | **0.1385** | **0.1828** | 0.3485 | 0.1811 |
| 1SAW | **0.1542** | **0.1936** | **0.1609** | **0.1942** | 0.2709 | 0.1859 |
| 1SIV | **0.1206** | **0.1728** | **0.1451** | **0.1692** | 0.1980 | 0.1479 |
| 1SPD | **0.1192** | **0.2021** | **0.1381** | **0.1942** | 0.2609 | 0.2034 |
| 1TOI | **0.1529** | **0.1871** | **0.1495** | **0.1826** | 0.3155 | 0.1735 |
| 1VDW | **0.1883** | **0.1975** | **0.1981** | **0.2056** | 0.2886 | 0.1960 |
| 1X2I | **0.1432** | **0.1774** | **0.1290** | **0.1807** | 0.1920 | 0.1634 |
| 1XHF | **0.1022** | **0.1796** | **0.1028** | **0.1722** | 0.3616 | 0.1710 |
| 1Y4J | **0.1540** | **0.1623** | **0.1499** | **0.1608** | 0.3214 | 0.1562 |
| 2A9U | **0.2792** | **0.3074** | **0.2233** | **0.3091** | 0.3297 | 0.2848 |
| 2AXW | **0.1407** | **0.2391** | **0.1373** | **0.2523** | 0.3942 | 0.2033 |
| 2B8W | **0.1940** | **0.1981** | **0.1893** | **0.1956** | 0.4988 | 0.1846 |
| 2B9D | **0.1755** | **0.1822** | **0.1568** | **0.1719** | 0.2858 | 0.1694 |
| 2BQP | **0.1076** | **0.1550** | **0.1240** | **0.1586** | 0.3948 | 0.1478 |
| 2E1N | **0.1973** | **0.2385** | **0.2230** | **0.2444** | 0.4458 | 0.2309 |
| 2F8F | **0.2080** | **0.2147** | **0.1909** | **0.2129** | 0.2596 | 0.2081 |
| 2FF1 | **0.1526** | **0.1609** | **0.1563** | **0.1647** | 0.3698 | 0.1574 |
| 2H84 | **0.1647** | **0.1855** | **0.1648** | **0.1859** | 0.2323 | 0.1762 |
| 2HS1 | **0.1575** | **0.1960** | **0.1526** | **0.1893** | 0.2093 | 0.1631 |

(*Continued*)



Published by Woodhead Publishing Limited, 2012

**Table 6.1** (*Continued*)

| PDB ID | *O/T* (1) | *O/R* (1) | *O/T* (2) | *O/R* (2) | *O/T* (dimer) | *O/R* (dimer) |
|---|---|---|---|---|---|---|
| 2NZH | **0.1771** | **0.2010** | **0.1452** | **0.2024** | 0.2078 | 0.1915 |
| 2Q2H | **0.1249** | **0.1802** | **0.1260** | **0.1666** | 0.1909 | 0.1666 |
| 2Q2I | **0.1234** | **0.1774** | **0.1307** | **0.1736** | 0.2275 | 0.1662 |
| 2Q3X | **0.1587** | **0.1972** | **0.1399** | **0.1949** | 0.2628 | 0.1936 |
| 2QX0 | **0.1236** | **0.1828** | **0.1263** | **0.1793** | 0.3860 | 0.1615 |
| 2Z5E | **0.1664** | **0.1845** | **0.1551** | **0.1968** | 0.1930 | 0.1758 |
| 3CNK | **0.1258** | **0.1561** | **0.1232** | **0.1536** | 0.1749 | 0.1533 |
| 3CRN | **0.1857** | **0.2201** | **0.1527** | **0.2248** | 0.4411 | 0.2239 |
| 3D57 | **0.1767** | **0.1797** | **0.1751** | **0.1764** | 0.4826 | 0.1729 |
| 3FY5 | **0.0684** | **0.1776** | **0.0586** | **0.1529** | 0.2716 | 0.1610 |
| 3FYF | **0.1349** | **0.1846** | **0.1765** | **0.2002** | 0.2750 | 0.1940 |
| 3GWL | **0.1465** | **0.1684** | **0.1516** | **0.1637** | 0.3026 | 0.1544 |
| 3GWN | **0.1303** | **0.1646** | **0.1414** | **0.1614** | 0.1811 | 0.1619 |
| One monomer accordant, dimer not accordant | | | | | | |
| 1CD0 | 0.1600 | 0.1315 | **0.1168** | **0.1442** | 0.3352 | 0.1358 |
| 1G57 | 0.1712 | 0.1602 | **0.1499** | **0.1674** | 0.2023 | 0.1611 |
| 1OH0 | **0.1276** | **0.1306** | 0.1549 | 0.1443 | 0.3002 | 0.1377 |
| 1R1V | 0.2630 | 0.1630 | **0.1460** | **0.1624** | 0.2849 | 0.1612 |
| 1RV3 | 0.2742 | 0.1703 | **0.1732** | **0.1748** | 0.3460 | 0.1647 |
| 1UEH | 0.1868 | 0.1734 | **0.1730** | **0.1770** | 0.3216 | 0.1630 |
| 1V8H | **0.1564** | **0.1610** | 0.1862 | 0.1532 | 0.2700 | 0.1594 |
| 2CD0 | **0.1293** | **0.1332** | 0.1348 | 0.1342 | 0.3440 | 0.1295 |
| 2J6W | **0.1452** | **0.1592** | 0.1673 | 0.1580 | 0.5450 | 0.1541 |
| 2NN2 | **0.1384** | **0.1507** | 0.1655 | 0.1548 | 0.3016 | 0.1483 |
| 3I4S | **0.1385** | **0.1557** | 0.1500 | 0.1453 | 0.1887 | 0.1480 |
| 3SDH | 0.1944 | 0.1675 | **0.1422** | **0.1582** | 0.2968 | 0.1628 |
| Both monomers not accordant, dimer accordant | | | | | | |
| 1CMB | 0.3326 | 0.2011 | 0.2835 | 0.2042 | **0.2129** | **0.2145** |
| 1F4N | 0.2381 | 0.1713 | 0.2518 | 0.1708 | **0.1787** | **0.1882** |
| 1Y7Q | 0.3144 | 0.2540 | 0.2987 | 0.2819 | **0.2304** | **0.3168** |



Published by Woodhead Publishing Limited, 2012

A departure from the idealized distribution of hydrophobicity can manifest itself as either a local deficiency or a local excess. The former case usually applies to empty binding pockets capable of housing ligands, whereas the latter case (if it occurs on the molecular surface) usually indicates a protein complexation site, where contact between protein molecules can eliminate the entropically disadvantageous proximity to water. For this reason, our search for binding pockets and compensation sites relied on locating the local maxima and minima of the $\Delta\tilde{H}$ profiles.

## 6.2 ROC curves

The receiver operating characteristic (ROC) curve method,[3] described in Chapter 5, was used to measure the accordance of the selected homodimers with the complexation model. Computations were performed separately for $\Delta\tilde{H}_i \leq 0$ and $\Delta\tilde{H}_i \geq 0$. The former case describes protein–protein complexation mediated by residues with excess hydrophobicity, whereas the latter corresponds to local hydrophobicity deficiencies. Table 6.1 shows how various proteins conformed to the model presented here, divided according to the degree of accordance with respect to the *O*/*T* (observed versus theoretical) and *O*/*R* (observed versus random) values for monomers and dimers.

The accordance of the hydrophobicity distribution with the assumed model can be measured by quantitative comparison of the idealized and observed hydrophobicity distributions throughout the protein body, presented in the form of a profile. To enable a proper interpretation of the distance between the theoretical and observed distributions, the corresponding distance between the



Published by Woodhead Publishing Limited, 2012

observed and random distributions must also be calculated. If the latter is greater than the former, the protein can be said to possess a 3D Gaussian-like hydrophobic core. The theoretical and random distributions are treated as a reference (target) against which the observed distribution can be analyzed. A quantitative measure of the agreement between the observed and idealized distributions or between the observed and random distributions is given by the Kullback–Leibler divergence entropy. The $O/T$ and $O/R$ values can be used to determine the specific type of distribution observed. $O/T < O/R$ means that the observed distribution approximates the theoretical values, whereas the reverse relation ($O/T > O/R$) indicates that the observed distribution is random in character.

The profile of $\Delta\tilde{H}$ (the difference between the theoretical and observed distributions) reveals the hydrophobicity differences for each residue in the polypeptide chain. The profile can be interpreted as follows:

1. Residues associated with higher-than-expected hydrophobicity are assumed to be responsible for protein complexation. A localized excess of hydrophobicity (minima in the $\Delta\tilde{H}_i$ profile), if occurring on the surface, is treated as a potential site of complexation driven by hydrophobic interactions.
2. Residues associated with hydrophobicity deficiencies (peaks in the $\Delta\tilde{H}_i$ profile) are assumed to be capable of interaction with external objects (ligands or other proteins). In such cases, the interacting molecule compensates the hydrophobicity deficiency occurring on the surface of the target protein.

In order to quantitatively measure these effects, ROC curves were additionally plotted, and the area between the diagonal and the corresponding ROC curve was treated as a quantitative measurement of the accordance between the model and the observations. A negative area was attributed to any ROC curve that lay below the diagonal; this was interpreted as a lack of accordance with the assumed model, similarly to the case of low positive values. The TPR (true positive rate) and FPR (false positive rate) were calculated for the $\Delta\tilde{H}_i$ values in a stepwise fashion. This procedure was performed separately for $\Delta\tilde{H}_i \geq 0$ and $\Delta\tilde{H}_i \leq 0$ to derive the characteristics of the residues associated with hydrophobicity deficiencies, and, as a consequence, distinguish between the two complexation models.

Table 6.2 presents a set of sample proteins, along with their ROC characteristics. For each protein, the accordance between its empirically observed hydrophobicity distribution and the idealized model (3D Gaussian function), measured using ROC curves and the area between the diagonal and the ROC curve, is listed. The dimers are grouped according to the classification described above based on the Kullback–Leibler entropy scale. A summary of the properties of the proteins taking the Kullback–Leibler entropy and ROC analysis into account is presented in Table 6.3. Proteins selected to represent each group are discussed below.

In addition to determining the structural accordance of each protein's hydrophobic core with the "fuzzy oil drop" model, ROC curve analysis also enables identification of the residues involved in protein complexation. Such complexation can be mediated by residues with excess hydrophobicity or by residues with hydrophobicity deficiencies. One protein that appears to agree with the internal-hydrophobicity distribution model but does not follow the predicted protein–protein complexation model is the chain 3I4S.A.[4]



Published by Woodhead Publishing Limited, 2012

**Table 6.2** **Results of ROC curve analysis. ROC curves were plotted for residues with $\Delta\tilde{H}_i \leq 0$ (hydrophobicity excess) and, separately, for residues with $\Delta\tilde{H}_i \geq 0$ (hydrophobicity deficiency). A negative area was attributed to ROC curves that lay below the diagonal. A positive area of 0.3 or more was assumed to indicate good agreement between the theoretical model and the empirical observations (values given in bold)**

| PDB ID | Chain | Area for $\Delta\tilde{H}_i \geq 0$ | Contact for $\Delta\tilde{H}_i \geq 0$ | Area for $\Delta\tilde{H}_i \leq 0$ | Contact for $\Delta\tilde{H}_i \leq 0$ |
|---|---|---|---|---|---|
| Hydrophobic core accordant (Kullback–Leibler), ligation accordant (ROC) for $\Delta\tilde{H}_i \geq 0$ | | | | | |
| 1HDF | B | **0.511** | 0.250 | −0.124 | 0.750 |
| 1HTI | A | **0.372** | 0.394 | 0.041 | 0.606 |
| 1HTI | B | **0.370** | 0.303 | −0.048 | 0.697 |
| 2A9U | A | **0.431** | 0.441 | 0.206 | 0.559 |
| 2A9U | B | **0.449** | 0.406 | 0.077 | 0.594 |
| 2B9D | A | **0.409** | 0.625 | −0.697 | 0.375 |
| 2E1N | B | **0.425** | 0.211 | 0.212 | 0.789 |
| 3CRN | A | **0.580** | 0.231 | 0.153 | 0.769 |
| 3FYF | A | **0.433** | 0.636 | 0.014 | 0.364 |
| Hydrophobic core accordant (Kullback–Leibler), ligation accordant (ROC) for $\Delta\tilde{H}_i \leq 0$ | | | | | |
| 1BFT | A | −0.454 | 0.562 | **0.381** | 0.438 |
| 1BFT | B | −0.432 | 0.467 | **0.375** | 0.533 |
| 1K68 | A | −0.500 | 0.167 | **0.352** | 0.833 |
| 1RRG | A | −0.402 | 0.250 | **0.329** | 0.750 |
| 1RRG | B | −0.278 | 0.250 | **0.306** | 0.750 |
| 2E1N | A | −0.156 | 0.273 | **0.437** | 0.727 |
| 2NZH | A | −0.215 | 0.645 | **0.304** | 0.355 |
| 2WCI | A | −0.277 | 0.333 | **0.436** | 0.667 |
| Hydrophobic core not accordant (Kullback–Leibler), ligation accordant (ROC) for $\Delta\tilde{H}_i \geq 0$ | | | | | |
| 1DLT | A | **0.338** | 0.387 | 0.205 | 0.613 |



Published by Woodhead Publishing Limited, 2012

| | | | | | |
|---|---|---|---|---|---|
| 1FZV | A | **0.430** | 0.429 | −0.238 | 0.571 |
| 1FZV | B | **0.382** | 0.357 | −0.132 | 0.643 |
| 1G8M | A | **0.588** | 0.755 | −0.111 | 0.245 |
| 1G8M | B | **0.532** | 0.780 | −0.077 | 0.220 |
| 1HUX | B | **0.467** | 0.300 | 0.220 | 0.700 |
| 1SD6 | A | **0.376** | 0.375 | 0.203 | 0.625 |
| 1SD6 | B | **0.348** | 0.536 | 0.288 | 0.464 |
| 1TR8 | A | **0.742** | 0.923 | −0.421 | 0.077 |
| 1TR8 | B | **0.714** | 1.000 | 0.000 | 0.000 |
| 1U3W | A | **0.319** | 0.353 | 0.179 | 0.647 |
| 1USP | A | **0.301** | 0.593 | −0.184 | 0.407 |
| 1V58 | A | **0.441** | 0.694 | −0.156 | 0.306 |
| 1V58 | B | **0.374** | 0.765 | 0.045 | 0.235 |
| 2ARV | A | **0.561** | 0.360 | 0.235 | 0.640 |
| 2G67 | A | **0.392** | 0.611 | −0.150 | 0.389 |
| 2R52 | B | **0.543** | 0.542 | −0.123 | 0.458 |
| Hydrophobic core not accordant (Kullback–Leibler), ligation accordant (ROC) for $\Delta\tilde{H}_i \leq 0$ | | | | | |
| 1G85 | A | 0.196 | 0.435 | **0.533** | 0.565 |
| 1G85 | B | 0.148 | 0.413 | **0.373** | 0.587 |
| 1HQO | A | 0.037 | 0.500 | **0.342** | 0.500 |
| 1L2U | B | −0.284 | 0.486 | **0.311** | 0.514 |
| 1NXH | B | 0.270 | 0.483 | **0.341** | 0.517 |
| 1SD4 | A | 0.251 | 0.472 | **0.368** | 0.528 |
| 1T09 | A | 0.160 | 0.442 | **0.471** | 0.558 |
| 1T09 | B | 0.042 | 0.340 | **0.453** | 0.660 |
| 1U3W | B | 0.037 | 0.472 | **0.317** | 0.528 |
| 1XJJ | B | −0.006 | 0.217 | **0.361** | 0.783 |
| 1Y7Q | A | 0.041 | 0.342 | **0.355** | 0.658 |
| 1YGA | A | −0.645 | 0.333 | **0.780** | 0.667 |
| 1YGA | B | −0.800 | 0.200 | **0.820** | 0.800 |
| 1YZ5 | A | −0.214 | 0.227 | **0.313** | 0.773 |
| 2FJT | A | −0.315 | 0.450 | **0.427** | 0.550 |
| 2FJT | B | −0.234 | 0.381 | **0.425** | 0.619 |

*(Continued)*



Published by Woodhead Publishing Limited, 2012

**Table 6.2** ***(Continued)***

| PDB ID | Chain | Area for $\Delta\tilde{H}_i \geq 0$ | Contact for $\Delta\tilde{H}_i \geq 0$ | Area for $\Delta\tilde{H}_i \leq 0$ | Contact for $\Delta\tilde{H}_i \leq 0$ |
|---|---|---|---|---|---|
| 2HA2 | B | −0.409 | 0.211 | **0.348** | 0.789 |
| 2QM8 | A | −0.164 | 0.463 | **0.505** | 0.537 |
| 2QM8 | B | −0.153 | 0.537 | **0.387** | 0.463 |
| Hydrophobic core not accordant (Kullback–Leibler), ligation accordant (ROC) for $\Delta\tilde{H}_i \geq 0$ and $\Delta\tilde{H}_i \leq 0$ | | | | | |
| 1HUX | A | **0.457** | 0.333 | **0.308** | 0.667 |
| 1NXH | A | **0.306** | 0.545 | **0.336** | 0.455 |
| 2ARV | B | **0.301** | 0.423 | **0.405** | 0.577 |
| 3GYZ | A | **0.527** | 0.435 | **0.560** | 0.565 |

1K68 is another protein in which the hydrophobic core appears to match theoretical predictions.[5] The Kullback–Leibler relation between the *O/T* and *O/R* ratios suggests good agreement between the empirical and expected values. This is evidenced in Figure 6.1 and Table 6.2.

Residues suspected of involvement in protein–protein interaction are associated with local minima of $\Delta\tilde{H}_i$ (excess hydrophobicity). The localization and characteristics of the $\Delta\tilde{H}_i$ values for the residues actually engaged in complexation are shown in Figure 6.1. According to this profile, almost all residues engaged in complexation satisfy $\Delta\tilde{H}_i \leq 0$. The local minimum of $\Delta\tilde{H}_i$ at position 120 shows particularly good agreement with the model. Only one residue engaged in complexation has $\Delta\tilde{H}_i \geq 0$. The automatic recognition of this case is shown in Figure 6.1. ROC curves plotted for residues with $\Delta\tilde{H}_i \leq 0$ and, separately, for residues with $\Delta\tilde{H}_i \geq 0$ reveal a greater involvement of residues with negative $\Delta\tilde{H}_i$ in complexation. The values representing the area between the ROC curves and the



Published by Woodhead Publishing Limited, 2012

**Table 6.3** Classification of proteins according to two criteria: (1) the Kullback–Leibler distance entropy (revealing proteins whose hydrophobic cores are structurally in accordance with the "fuzzy oil drop" model), and (2) analysis of the ROC curves (revealing proteins whose complexation areas are structurally in accordance with the "fuzzy oil drop" model). The underlined codes indicate proteins whose dimeric forms are also in accordance with the model (using the *O/T* and *O/R* criterion). The protein 1MB4 was selected as an example to represent discordance with respect to the two techniques applied

| | | ROC curve analysis | | |
|---|---|---|---|---|
| | | Accordant | | Not accordant |
| | | $\Delta\tilde{H}_i > 0$ | $\Delta\tilde{H}_i < 0$ | |
| Divergence entropy | Accordant | 1F37, 1HDF.B, 1HTI.AB, 2A9U.AB, 2B9D.A, 2E1N.B, 2WCI, 3CRN.A, 3FYF.A | 1BFT.AB, 1K68.A, 1RRG.AB, 2E1N.A, 2NZH.A, 2Q3A.B, 2WCI.A | 1AC6.AB, 1B8Z.AB, 1BEB.AB, 1BFM.AB, 1B04.AB, 1C9O.AB, 1CDO.B, 1CMB, 1F05.AB, 1F37.AB, 1F4N, 1F9Z.AB, 1GOH.AB, 1G57.B, 1HDF.A, 1JUB.AB, 1K66.AB, 1K68.B, 1MY7.AB, 1OHO.A, 1QLL.AB, 1R1V.B, 1RB7.AB, 1RV3.B, 1SAW.AB, 1SIV.AB, 1SPD.AB, 1TOI.AB, 1UEH.B, 1V8H.A, 1VDW.AB, 1X2I.AB, 1XHF.AB, 1Y4J.AB, 1Y7Q, 1YE5.AB, 2AXW.AB, 2B8W.AB, 2B9D.B, 2BQP.AB, 2CDO.A, 2F8F.AB, 2FF1.AB, 2H84.AB, 2HS1.AB, 2J6W.A, 2NN2.A, 2NZH.B, 2Q2H.AB, 2Q2I.AB, 2Q3A.A, 2Q3X.AB, 2QXO.AB, 2VOC.AB, 2WCI.B, 2Z5E.AB, 3CNK.AB, 3CRN.B, 3D57.AB, 3FY5.AB, 3FYF.B, 3GWL.AB, 3GWN.AB, 3I4S.A, 3SDH.B |
| | Not accordant | 1AC6, 1AOZ, 1BEB, 1BFT, 1C3B, 1DLG, 1DLT.A, 1DV1, 1EK6, 1EXV, 1F05, 1FZV.AB, 1GOH, 1G8M.AB, 1HDF, 1HQO, 1HUX.AB, 1K66, 1K68, 1KXR, 1N4O, 1NOW, 1NXH.A, 1PSC, 1QHM, 1QLL, 1R1V, 1SAW, 1SD6.AB, 1SMS, 1SQI, 1TK3, 1TR8.AB, 1U3W.A, 1UEH, 1USP.A, 1V58.AB, 1VDW, 1XJJ, 1XW5, 2AA4, 2ARV.AB, 2B9D, 2BQP, 2DQ3, 2E4U, 2ERO, 2F8F, 2FA2, 2G67.A, 2HA2, 2J6W, 2OP9, 2P4Y, 2PFL, 2PZE, 2Q2O, 2Q3X, 2QXO, 2R52.B, 3D57, 3DUP, 3FY5, 3GYZ.A, 3SDH | 1G85.AB, 1HQO.A, 1HUX.A, 1L2U.B, 1NXH.AB, 1SD4.A, 1T09.AB, 1U3W.B, 1XJJ.B, 1Y7Q.A, 1YGA.AB, 1YZ5.A, 2ARV.B, 2FJT.AB, 2HA2.B, 2QM8.AB, 3GYZ.A | 1MB4.B taken as example, other proteins belonging to this group not discussed in this chapter |



Published by Woodhead Publishing Limited, 2012

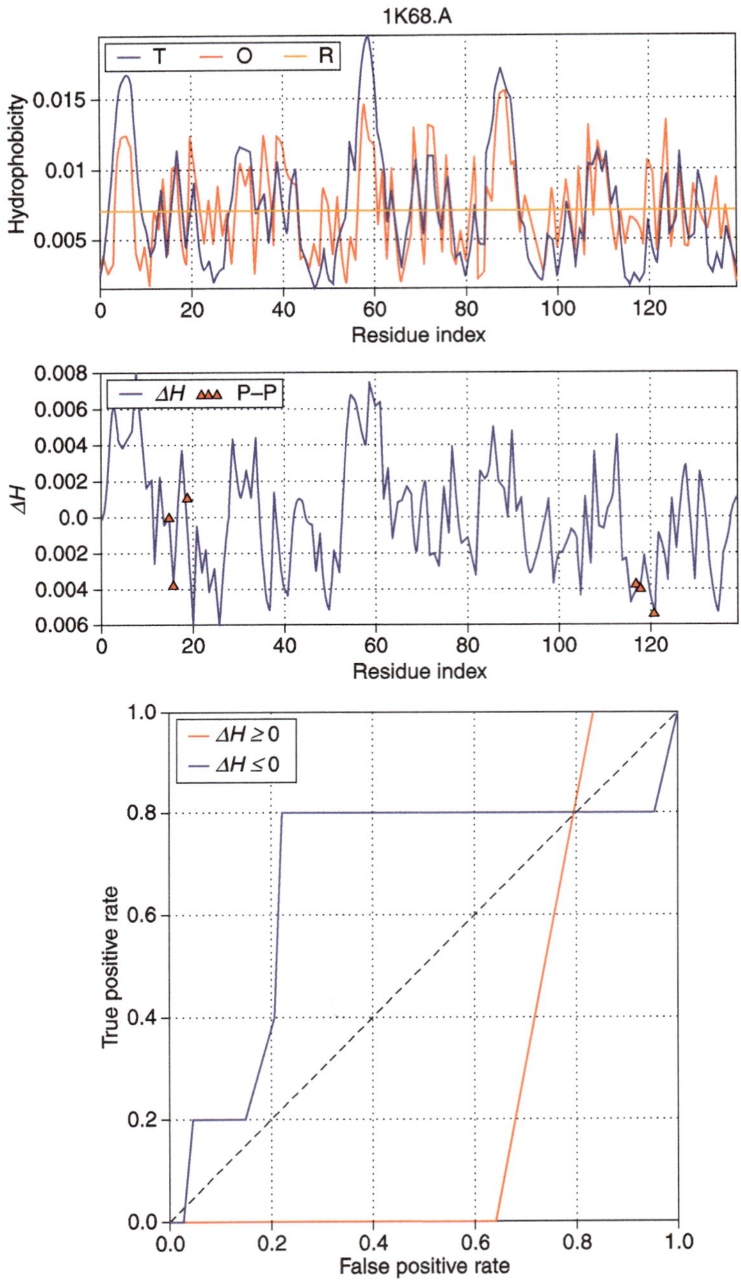


**Figure 6.1** Caption opposite



Published by Woodhead Publishing Limited, 2012

corresponding diagonal seem to support this observation (Table 6.3). The identification of the residues responsible for complexation in both cases ($\Delta\tilde{H}_i \leq 0$ and $\Delta\tilde{H}_i \geq 0$) followed a similar approach and produced two discrete sets of residues. The protein–protein complexation model (based on hydrophobic interactions) presented here seems to be accurately reflected by the results for 1K68, as well as by the results for the proteins highlighted in Table 6.2 as being in accordance with both of the criteria (i.e., $\Delta\tilde{H}_i \leq 0$).

The chain 3FYF.A was selected as an example of a hydrophobic core that is structurally in accordance with the "fuzzy oil drop" model. The relation between *O/T* and *O/R* for it (Table 6.2) supports this observation.[6] In this protein,

**Figure 6.1** Analysis of the chain 1K68.A, showing good accordance with respect to the hydrophobicity distribution as well as the identification of residues engaged in protein-protein complexation. Top: theoretical (blue) and observed (red) hydrophobicity distributions compared with random distribution (orange). According to the Kullback-Leibler distance entropy model, $O/T < O/R$, which suggests good agreement with theoretical predictions. Center: $\Delta\tilde{H}_i$ profile (blue), with residues engaged in protein-protein complexation marked in orange. Bottom: ROC curves. The blue line represents excess hydrophobicity (treated as a criterion of identification), and the red line represents residues with hydrophobicity deficiencies. The relatively large area between the diagonal and the blue ROC curve suggests good accordance with the protein-protein complexation model



Published by Woodhead Publishing Limited, 2012

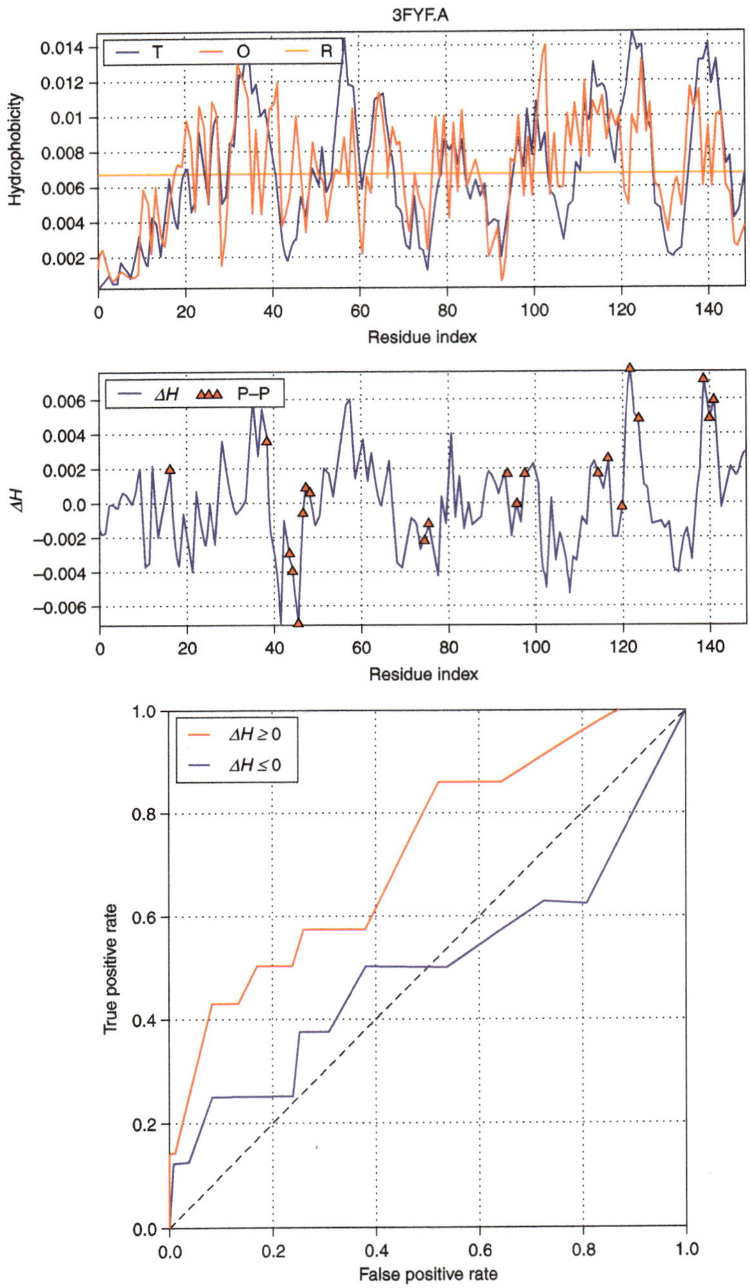


**Figure 6.2** Caption opposite



Published by Woodhead Publishing Limited, 2012

all residues engaged in complexation satisfy $\Delta\tilde{H}_i \geq 0$ (Figure 6.2). ROC curve analysis (Table 6.3) reveals a greater involvement of residues in dimerization (Figure 6.2). This protein (as well as other similar examples listed in Table 6.2) matches the proposed "mutual compensation" model. In two cases, the dimeric forms are also structurally in accordance with the "fuzzy oil drop" model, lending further support to this interpretation.

Another protein that exhibits good agreement with theoretical predictions regarding the structure of its hydrophobic core is 3I4S.[4] The degree of accordance between the expected and observed distributions can be seen in

**Figure 6.2** Analysis of the chain 3FYF.A, an example of accordance with respect to the hydrophobicity distribution as well as the identification of residues (with a hydrophobicity deficiency) engaged in protein–protein complexation. Top: theoretical (blue) and observed (red) hydrophobicity distributions compared with random distribution (orange). According to the Kullback–Leibler distance entropy, $O/T < O/R$, suggesting good agreement between the observed and idealized distributions. Center: $\Delta\tilde{H}_i$ profile (blue), with residues engaged in protein–protein complexation marked in orange. Bottom: ROC curves. The blue line corresponds to residues with excess hydrophobicity (treated as a criterion of identification), and the red line corresponds to residues associated with hydrophobicity deficiencies. The relatively large area between the diagonal and the ROC curve representing "deficient" residues validates the prediction that such residues are actually involved in protein–protein complexation



Published by Woodhead Publishing Limited, 2012

Figure 6.3. In this case, no specific characteristics can be attributed to the residues responsible for complexation: the two ROC curves appear similar in shape, making it impossible to accurately identify complexation sites (see Figure 6.3 again).

The protein 2FJT does not follow theoretical predictions with respect to the structure of its hydrophobic core.[7] Moreover, the observed and expected hydrophobicity distributions appear to differ significantly (Figure 6.4). This is further supported by the relation between *O/T* and *O/R*, although most of the residues involved in complexation seem to satisfy $\Delta\tilde{H}_i \leq 0$ (Figure 6.4 and Table 6.3).

**Figure 6.3** Analysis of the chain 3I4S.A, an example of good accordance with respect to the hydrophobicity distribution and poor accordance with respect to the identification of residues engaged in protein-protein complexation (no distinguishing properties). Top: theoretical (blue) and observed (red) hydrophobicity distributions compared with random distribution (orange). According to the Kullback-Leibler distance entropy, $O/T < O/R$, suggesting good agreement between the observed and idealized distributions. Center: $\Delta\tilde{H}_i$ profile (blue), with residues engaged in protein-protein complexation marked in orange. Bottom: ROC curves. The blue line corresponds to residues with excess hydrophobicity (treated as a criterion of identification), and the red line corresponds to residues associated with hydrophobicity deficiencies. As the two curves are similar in shape, residues responsible for complexation cannot be distinguished



Published by Woodhead Publishing Limited, 2012

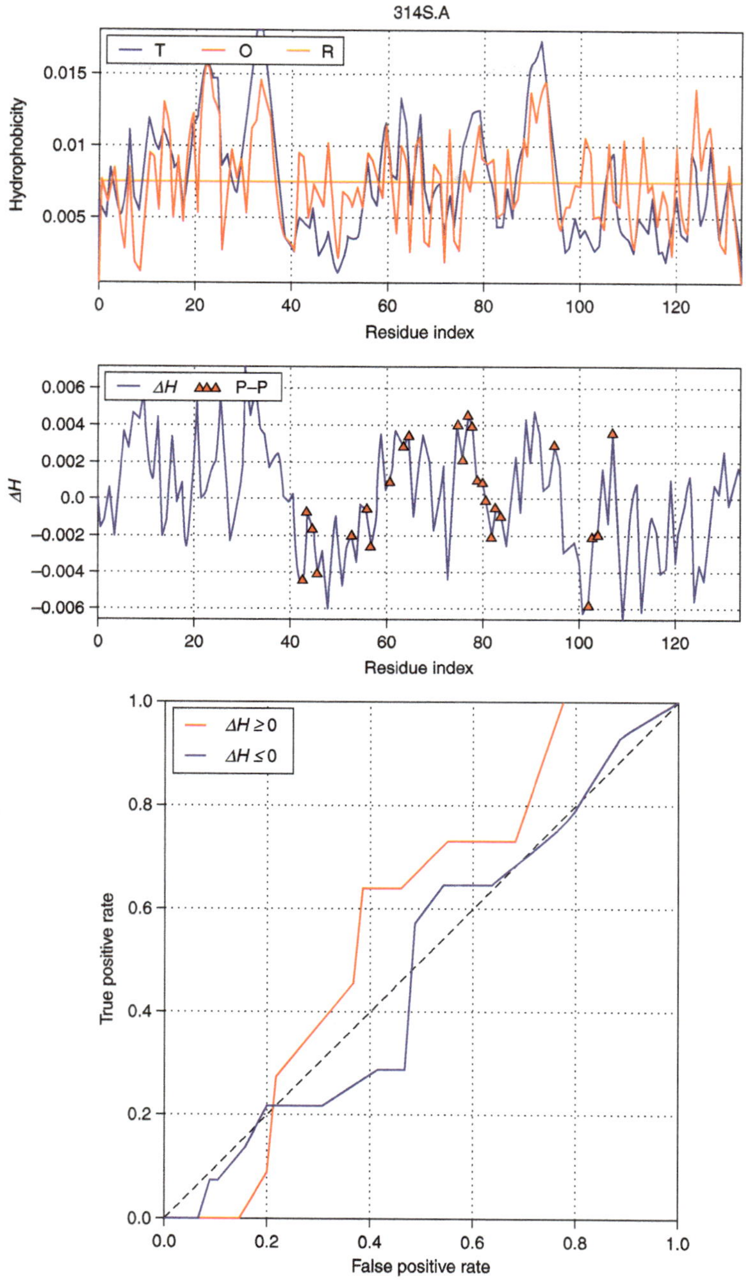


**Figure 6.3** Caption opposite



Published by Woodhead Publishing Limited, 2012

The next example showcases poor agreement with respect to the hydrophobicity distribution, along with a clearly discernible involvement of residues with $\Delta\tilde{H}_i \geq 0$ in protein complexation: this is demonstrated by the protein labeled 1TR8.[8] The Kullback–Leibler characteristics and the graphic depiction of the $\Delta\tilde{H}_i$ profile suggest that this protein diverges from the assumed model (Table 6.2 and Figure 6.5), but, on the other hand, ROC curve analysis indicates good accordance. Residues with $\Delta\tilde{H}_i \geq 0$ seem to be engaged in complexation (Figure 6.5 and Tables 6.2 and 6.3).

**Figure 6.4** Analysis of the chain 2FJT.B, an example of poor accordance with respect to the hydrophobicity distribution as well as the identification of residues (excess hydrophobicity) engaged in protein–protein complexation. Top: theoretical (blue) and observed (red) hydrophobicity distributions compared with random distribution (orange). According to the Kullback–Leibler distance entropy, *O/T* > *O/R*, suggesting poor agreement between the observed and idealized distributions of hydrophobicity. Center: $\Delta\tilde{H}_i$ profile (blue), with residues engaged in protein–protein complexation marked in orange. Bottom: ROC curves. The blue line corresponds to residues with excess hydrophobicity (treated as a criterion of identification), and the red line corresponds to residues associated with hydrophobicity deficiencies. The relatively large area between the diagonal and the ROC curve representing residues with excess hydrophobicity suggests good applicability of the theoretical model, in which such residues are suspected of involvement in protein–protein complexation



Published by Woodhead Publishing Limited, 2012

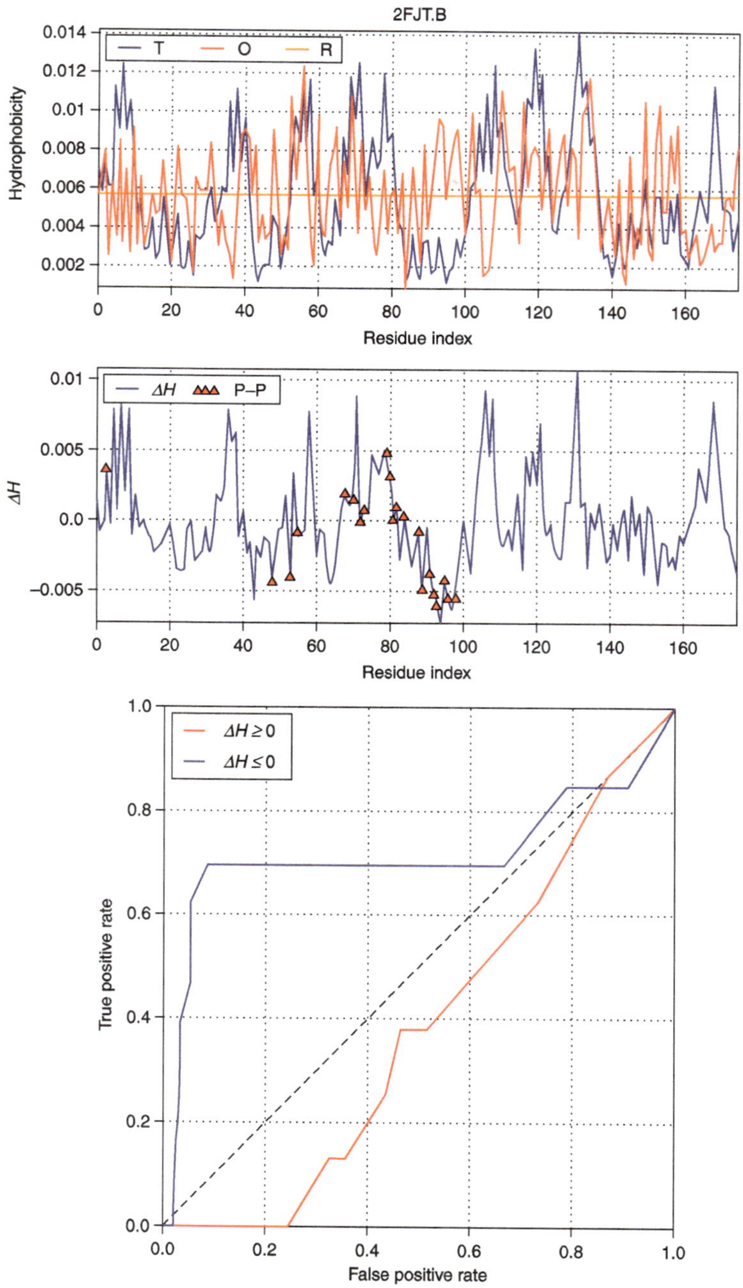


**Figure 6.4** Caption opposite



Published by Woodhead Publishing Limited, 2012

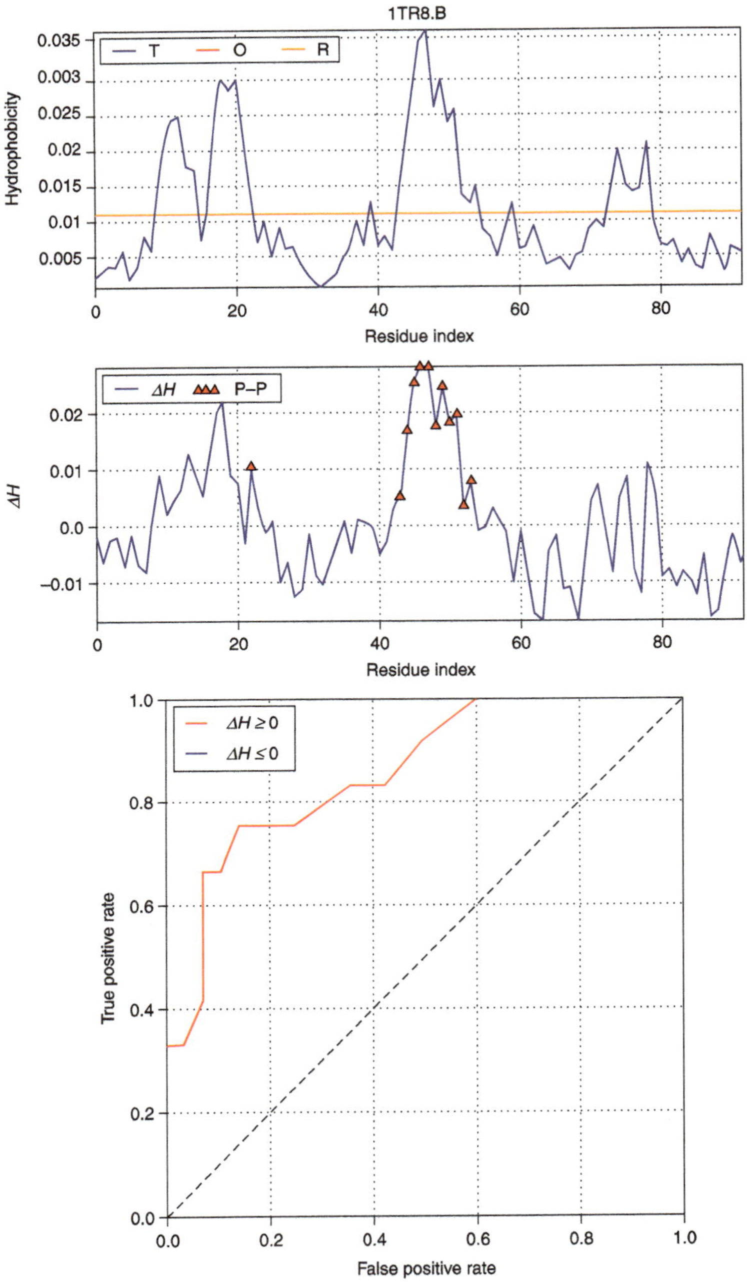


**Figure 6.5** **Caption opposite**



Published by Woodhead Publishing Limited, 2012

The final example selected for detailed analysis was the 1MB4 dimer, where the hydrophobic core is in accordance with the assumed model but no accordance with the complexation model is evident (Figure 6.6 and Tables 6.2 and 6.3).[9]

## 6.3 Conclusions

The heuristic "fuzzy oil drop" model is thought to enable *in silico* simulations of protein folding through optimization

**Figure 6.5** Analysis of the chain 1TR8.B, an example of poor accordance with respect to the hydrophobicity distribution and good accordance with respect to the identification of residues (with a hydrophobicity deficiency) involved in protein–protein complexation. Top: theoretical (blue) and observed (red) hydrophobicity distributions compared with random distribution (orange). According to the Kullback-Leibler distance entropy, $O/T > O/R$, suggesting poor agreement between the observed and idealized hydrophobicity distributions. Center: $\Delta\tilde{H}_i$ profile (blue), with residues involved in protein–protein complexation marked in orange. Bottom: ROC curve. The relatively large area between the diagonal and the ROC curve representing "deficient" residues suggests good applicability of the theoretical model, in which such residues are suspected of involvement in protein–protein complexation. Only one line (red) is actually drawn, as only those residues which are associated with local $\Delta\tilde{H}_i$ maxima are involved in complexation



Published by Woodhead Publishing Limited, 2012

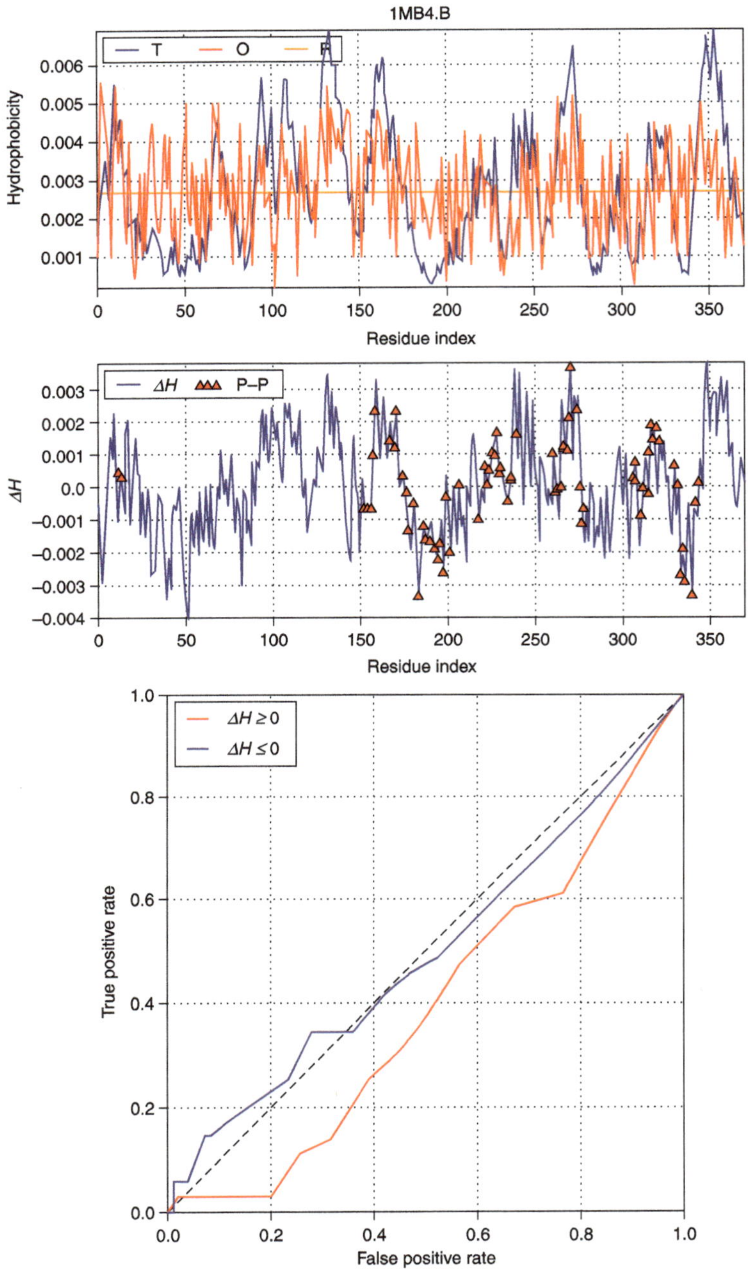


**Figure 6.6** **Caption opposite**



Published by Woodhead Publishing Limited, 2012

of nonbonding interactions. As far as hydrophobic interactions are concerned, we need to take account of internal optimization of the hydrophobic forces between the individual residues and also the influence of the external environment (the solvent water), which directs hydrophobic residues towards the center of the protein body and leads to exposure of hydrophilic residues on the protein surface. Extending the optimization procedure to cover hydrophobic/hydrophilic relations with external molecules reflects the influence of the protein's environment. Proteins fully in accordance with the idealized "fuzzy oil drop" model are characterized by excellent solubility but lack biological activity, since an "ideal" (hydrophilic) protein surface always prefers contact with water to interaction with other molecules. This is why irregularities in the distribution of hydrophobicity

**Figure 6.6** **Analysis of the chain 1MB4.B, an example of poor accordance with respect to the hydrophobicity distribution and with respect to the identification of residues involved in protein complexation. Top: theoretical (blue) and observed (red) hydrophobicity distributions compared with random distribution (orange). According to the Kullback–Leibler distance entropy, $O/T > O/R$, suggesting poor agreement between the observed and idealized hydrophobicity distributions. Center: $\Delta\tilde{H}_i$ profile (blue), with residues engaged in protein–protein complexation marked in orange. Bottom: ROC curves. The blue line corresponds to residues with excess hydrophobicity (treated as a criterion of identification), and the red line corresponds to residues associated with hydrophobicity deficiencies. As the two curves are similar in shape, residues responsible for complexation cannot be distinguished**



Published by Woodhead Publishing Limited, 2012

seem to correspond to specific biological profiles of proteins, particularly in the context of binding ligands.[10]

The "fuzzy oil drop" model has also been applied to *in silico* predictions of protein complexation, based on the results of homodimer analysis presented in this chapter. The primary conclusion is that many different, unrelated mechanisms are in fact responsible for complexation. Analysis of the results presented in Table 6.1 suggests at least two *in silico* complexation models. Note that the hydrophobicity profile may differ between the two monomers (although not to a significant degree) – this phenomenon may provide hints as to the sequential nature of the folding and complexation processes.

An interesting conclusion is that some proteins appear to be in accordance with the "fuzzy oil drop" model with respect to complexation areas without exhibiting similar accordance with respect to the Kullback–Leibler criterion. Such proteins are usually based on long polypeptide chains and often consist of multiple domains. Domain-selective accordance with hydrophobicity distribution models will be a subject of further study.

Proteins that contain SS bonds are of special interest. Their structural accordance with the "fuzzy oil drop" model and the presence of numerous SS bonds (for example in 1QLL) points to a possible mechanism for the generation of SS bonds.[11] It is possible that a "collaboration" of sorts develops between the hydrophobic core and the SS bond generation mechanism. The discovery of molecules with eight SS bonds (2E4U),[12] as well as proteins with three internal and two interchain SS bonds (1FZV),[13] raises the question of the nature and importance of such relations. We have noted the relatively large subset of proteins with SS bonds in our study group, and we intend to investigate this issue further.

In summary, we may conclude that the search for protein complexation prediction methods must take account of



Published by Woodhead Publishing Limited, 2012

various models and mechanisms that enable the generation of multiprotein complexes. The CAPRI (Critical Assessment of PRediction of Interactions)[14] initiative is a particularly noteworthy contribution to the field, as it tracks progress in the development of new algorithms for locating complexation sites and areas of intermolecular contact.[15–25]

## 6.4 References

1. Needleman S., Wunsch C. (1970) A general method applicable to the search for similarities in the amino acid sequence of two proteins. *J. Mol. Biol.* 48, 443–53.
2. Protein Data Bank. Available from: *http://www.rcsb.org/pdb/home/home.do* [accessed 7 May 2012].
3. Fawcett T. (2006) An introduction to ROC analysis. *Pattern Recognit. Lett.* 27, 861–74.
4. Patskovsky Y., Ramagopal U., Toro R., Freeman J., Do J., et al. Crystal structure of histidine triad protein from *Bradyrhizobium japonicum*. Available from: *http://www.rcsb.org/pdb/explore/explore.do?structureId=3i4s* [accessed 7 May 2012].
5. Benda C., Scheufler C., Tandeau de Marsac N., Gaertner W. (2004) Crystal structures of two cyanobacterial response regulators in apo- and phosphorylated form reveal a novel dimerization motif of phytochrome-associated response regulators. *Biophys. J.* 87, 476–87.
6. Patskovsky Y., Bonnano J.B., Ozyurt S., Rutter M., Chang S., et al. Crystal structure of protein Bvu-3222 from *Bacteroides Vulgatus*. Available from: *http://www.rcsb.org/pdb/explore/explore.do?structureId=3fyf* [accessed 7 May 2012].
7. Gallagher D.T., Smith N.N., Kim S.-K., Heroux A., Robinson H., Reddy P.T. (2006) Structure of the class IV



Published by Woodhead Publishing Limited, 2012

adenylyl cyclase reveals a novel fold. *J. Mol. Biol.* 362, 114–22.
8. Spreter T., Pech M., Beatrix B. (2005) The crystal structure of archaeal nascent polypeptide-associated complex (NAC) reveals a unique fold and the presence of a ubiquitin-associated domain. *J. Biol. Chem.* 280, 15849–54.
9. Blanco J., Moore R.A., Kabaleeswaran V., Viola R.E. (2003) A structural basis for the mechanism of aspartate-beta-semialdehyde dehydrogenase from *Vibrio cholerae. Protein Sci.* 12, 27–33.
10. Roterman I., Konieczny L., Jurkowski W., Prymula K., Banach M. (2011) Two-intermediate model to characterize the structure of fast-folding proteins. *J. Theor. Biol.* 283(1), 60–70.
11. Lee W.H., Da Silva Giotto M.T., Marangoni S., Toyama M.H., Polikarpov I., Garratt R.C. (2001) Structural basis for low catalytic activity in Lys49 phospholipases A2 – a hypothesis: the crystal structure of piratoxin II complexed to fatty acid. *Biochemistry* 40, 28–36.
12. Muto T., Tsuchiya D., Morikawa K., Jingami H. (2007) Structures of the extracellular regions of the group II/III metabotropic glutamate receptors. *Proc. Natl. Acad. Sci. USA* 104, 3759–64.
13. Iyer S., Leonidas D.D., Swaminathan G.J., Maglione D., Battisti M., et al. (2001) The crystal structure of human placenta growth factor-1 (PlGF-1), an angiogenic protein, at 2.0 Å resolution. *J. Biol. Chem.* 276, 12153–61.
14. Janin J. (2010) Protein–protein docking tested in blind predictions: the CAPRI experiment. *Mol. Biosyst.* 6, 2351–62.
15. Janin J. (2007) The targets of CAPRI rounds 6–12. *Proteins* 69, 699–703.



Published by Woodhead Publishing Limited, 2012

16. Janin J., Wodak S. (2007) The third CAPRI assessment meeting Toronto, Canada, April 20–21, 2007. *Structure* 15, 755–9.
17. Janin J., Rodier F., Chakrabarti P., Bahadur R.P. (2007) Macromolecular recognition in the Protein Data Bank. *Acta Crystallogr. D: Biol. Crystallogr.* 63, 1–8.
18. Janin J. (2005) The targets of CAPRI rounds 3–5. *Proteins* 60, 170–5.
19. Janin J. (2005) Sailing the route from Gaeta, Italy, to CAPRI. *Proteins* 60, 149.
20. Janin J. (2005) Assessing predictions of protein–protein interaction: the CAPRI experiment. *Protein Sci.* 14, 278–83.
21. Janin J., Séraphin B. (2003) Genome-wide studies of protein–protein interaction. *Curr. Opin. Struct. Biol.* 13, 383–8.
22. Li C.H., Ma X.H., Chen W.Z., Wang C.X. (2003) A soft docking algorithm for predicting the structure of antibody–antigen complexes. *Proteins* 52, 47–50.
23. Janin J., Henrick K., Moult J., Eyck L.T., Sternberg M.J., et al. (2003) Critical Assessment of PRedicted Interactions. *Proteins* 52, 2–9.
24. Janin J. (2002) Welcome to CAPRI: a Critical Assessment of PRedicted Interactions. *Proteins* 47, 257.
25. Vajda S., Vakser I.A., Sternberg M.J.E., Janin J. (2002) Modelling of protein interactions in genomes. *Proteins* 47, 444–6.

## 6.5 Suggested reading

Inbar Y., Benyamini H., Nussinov R., Wolfson H.J. (2005) Combinatorial docking approach for structure prediction



Published by Woodhead Publishing Limited, 2012

of large proteins and multi-molecular assemblies. *Phys Biol.* 2(4), S156–65.

Keskin O., Ma B., Rogale K., Gunasekaran K., Nussinov R. (2005) Protein–protein interactions: organization, cooperativity and mapping in a bottom-up Systems Biology approach. *Phys. Biol.* 2(2), S24–35.

Ofran Y., Rost B. (2003) Analysing six types of protein–protein interfaces. *J. Mol. Biol.* 2003, 325(2), 377–87.

Rual J.F., Venkatesan K., Hao T., Hirozane-Kishikawa T., Dricot A., et al. (2005) Towards a proteome-scale map of the human protein–protein interaction network. *Nature* 437(7062), 1173–8.

Tarassov K., Messier V., Landry C.R., Radinovic S., Serna Molina M.M., et al. (2008) An in vivo map of the yeast protein interactome. *Science* 320(5882), 1465–70.



Published by Woodhead Publishing Limited, 2012

7

# Simulation of the polypeptide chain folding process using the "fuzzy oil drop" model

*L. Konieczny and I. Roterman-Konieczna*

**Abstract:** The proposed folding model (involving ES and LS intermediate stages) requires some additional work if it is to enable useful simulations. Specifically, it requires guidelines for reducing the volume of the bounding "drop" in a controlled manner and reconciling the optimization of the internal field and the external (hydrophobicity) field. The model also needs to be validated in the context of ligand binding and protein complexation with a suitable case study. Results produced by our algorithms indicate that the presence of external factors (such as ligand molecules) may directly affect the polypeptide chain folding process, ensuring the creation of a highly specific binding pocket while also preserving the "oil drop" hydrophobic core.

**Key words:** polypeptide chain folding process, influence of external environment, influence of aqueous environment, influence of external nonprotein molecules (ligands).



Published by Woodhead Publishing Limited, 2012

## 7.1 Simulation of the folding process in the presence of an external hydrophobic force field

Having presented the basic assumptions for the individual components of our proposed model, we now have to describe how these components fit together. Figure 7.1 presents the means of acquiring the necessary information about a protein molecule, and an algorithm that attempts to recreate its structure. The left-hand side of the image refers to the so-called "step-back" process (simulating the folding process in reverse), and the right-hand side corresponds to a folding simulation, exploiting the data produced by the "step-back" analysis.

The primary source of input for our simulation is the crystalline protein structure, as listed in the PDB repository (Figure 7.1(a)). The angles $\Phi$ and $\Psi$ are calculated for the polypeptides determining the conformation of the chain (Figure 7.1(b)). The "step-back" process is carried out by introducing the limited conformational space to a section (subspace) determined by the appropriate angles $\Phi_e$ and $\Psi_e$ while preserving the "least possible distance" criterion between each pair $\{\Phi, \Psi\}$ and the corresponding angles $\{\Phi_e, \Psi_e\}$ plotted on the elliptical path. The resulting structure is illustrated in Figure 7.1(c).

Performing this operation for a large (nonredundant) set of proteins results in a probability profile describes the stochastic distribution of "preferred" conformations within the limited conformational subspace for each amino acid (Figure 7.1(d)). By assigning letter codes to specific fragments of the elliptical path (Figure 7.1(d)), it is possible to express the structure of the entire chain (in its early-stage structural form) as a sequence of letters, each of which corresponds to a local maximum of the probability profile in the ES



Published by Woodhead Publishing Limited, 2012

intermediate (Figure 7.1(e)). Repeating this process for each protein (in the nonredundant PDB data set) in the study produces a contingency table that expresses the relation between the ES intermediate conformation and specific amino acid sequences. Our study relies on dividing the input chain into sequences consisting of four peptides: the tetrapeptide is the shortest possible fragment that can be unambiguously translated into a specific secondary structural motif.

Once fully assembled, the contingency table becomes a source of information regarding the preferred conformations, indicating which structural motifs are most frequently observed within the limited conformational subspace for each tetrapeptide.

The right-hand side of Figure 7.1 represents a simulation that produces a model of the protein's final structure. The sole input for this *in silico* folding process is the amino acid sequence itself. On the basis of the values contained in the contingency table, one can assign structural codes (as shown in Figure 7.1(d)) to each tetrapeptide found in the input chain, thereby obtaining a set of four character strings (this is due to partial overlaps between tetrapeptide sequences) (Figure 7.1(f)). From this set one can select a final "consensus" sequence, based on repeatability or weighing criteria.[1]

The assumed (consensus-based) early-stage intermediate structure hints at the preferred values of $\Phi_e$ and $\Psi_e$; however, the actual angles may not be exactly the same as those obtained in the "step-back" algorithm. This is because in each step only a single local-maximum position can be taken into account for a given letter of the structural alphabet (Figure 7.1(g)). As a result, the structure depicted in Figure 7.1(h) differs from the one generated by the "step-back" procedure (Figure 7.1(c)). In particular, Figure 7.1(f)



Published by Woodhead Publishing Limited, 2012

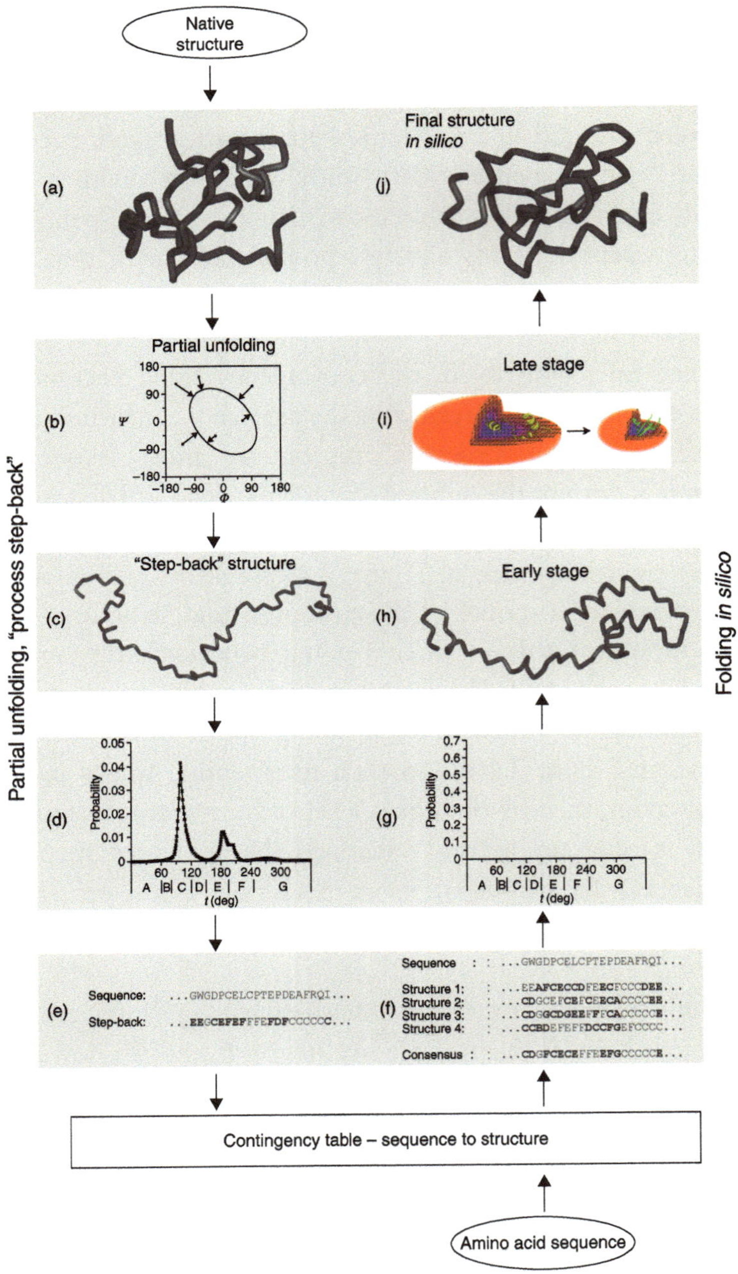


**Figure 7.1** **Caption opposite**



Published by Woodhead Publishing Limited, 2012

**Figure 7.1** Schematic depiction of the model presented here. The left-hand side of the figure shows how to obtain the necessary information regarding the early-stage intermediate (the "step-back" procedure – unfolding the polypeptide chain to study its initial form). The algorithm reads a PDB file (a) and proceeds to determine the angles $\Phi$ and $\Psi$ for each residue. These values are subsequently translated into the corresponding angles $\Phi_e$ and $\Psi_e$, which belong to the limited conformational subspace provided by the ES model (b). The structure, changed according to the change of $\Phi$ and $\Psi$ to $\Phi_e$ and $\Psi_e$ (the ES structural form), is shown in (c). Knowing the probability profile along the elliptical path for each amino acid enables us to express the structure of the polypeptide chain as a sequence of letters belonging to a predefined "structural alphabet" (d). This, in turn, means that the sequence and structure can both be expressed in a formal language suitable for processing with specialized tools for structural modeling and similarity assessment, such as ClustalW (e). The intermediate input data (the sequence of letters describing the ES intermediate) can be used to prepare a contingency table, where each tetrapeptide fragment is stochastically associated with specific structural motifs.

The right-hand side of the figure presents the polypeptide chain folding process, conducted in accordance with our model. The input data (a sequence of amino acids) is matched to structural codes describing the ES intermediate. Owing to partial overlaps between the individual tetrapeptide fragments, the procedure produces four separate strings of structural codes. The next step is therefore to find a consensus structure on the basis of the most frequently occurring structural motifs for each item in the sequence (f). Each structural code is associated with a local maximum in the probability profile – this gives the corresponding $\Phi_e$ and $\Psi_e$ values (g), which are used to generate the early-stage intermediate structure (h). Note that this result differs from the outcome of the "step-back" algorithm owing to the discretization performed in step (f). Further steps must take the presence of the water environment into account, expressed by a 3D Gaussian "drop" with a progressively shrinking volume (i). The final result of the process is depicted in (j)



Published by Woodhead Publishing Limited, 2012

suggests that the structure shown in Figure 7.1(h) may represent unacceptable collisions. In order to resolve possible collisions, a so-called "letter traversal" algorithm can be applied, which involves stepwise changes to $\Phi_e$ and $\Psi_e$, although these changes are allowed only for a particular fragment (i.e., a specific letter code). The resulting structure provides input for the late-stage folding simulation.

The molecule shown in Figure 7.1(h) is subsequently optimized with regard to internal forces (i.e., pairwise electrostatic, van der Waals, and torsion-potential interactions). Once this step is completed, the polypeptide is placed inside a "drop" that entirely encapsulates the folded chain (Figure 7.1(i)). This is followed by optimization of the hydrophobic interactions, where the criterion for accordance is the difference between the actual distribution of hydrophilic and hydrophobic residues (resulting from interactions between side chains) and the theoretical (idealized) distribution modeled by a 3D Gaussian function.

Interleaving the two modes of optimization (minimization of internal forces and minimization of the differences between the observed and theoretical hydrophobicity distributions) leads eventually to a consensus: the protein molecule converges to a state in which both types of interactions are reflected and the tertiary structure takes account of the indirect influence of water on the protein's hydrophobic core. Additionally, each round of optimization reduces the volume of the encapsulating "drop," which eventually approaches the optimal value for a polypeptide chain of a given length (specific volume limits can be derived from statistical data found in the PDB).[2]

It is necessary at this point to provide a more in-depth description of the step shown in Figure 7.1(i). The proposed model takes into account the external environment, which is modeled by a 3D Gaussian "fuzzy oil drop" function. This function reflects the influence of the aqueous solvent on the folding process, leading to the location of hydrophobic residues in the interior with simultaneous exposure of hydrophilic residues on the surface. The process occurs in tandem with optimization of the internal electrostatic, van der Waals, and torsion-potential forces, all of which are treated in a pairwise fashion. Note that the internal interactions tend to follow a random distribution, with a trend towards local optima and producing similar conditions throughout the protein body. In contrast, the interactions with the external environment must be treated as a global phenomenon (involving the entire molecule). The local optimization is performed in the traditional manner, by minimizing pairwise interaction forces, whereas the global optimization relies on minimizing the discrepancies between the observed structure and the corresponding (idealized) hydrophobic-core model, described by a 3D Gaussian function. In order to perform this operation accurately, it is necessary first to prepare an ES intermediate, based on the contingency table presented here. The structure of this intermediate is usually characterized by very loose packing (Figures 7.1(c) and (h)) and often assumes the form of an elongated strand. Consequently, the encapsulating "drop" is initially very large and devoid of any appreciable hydrophobicity spike at its center (the Gaussian law implies that the peak hydrophobicity value is inversely proportional to the volume of the drop, i.e., its coefficients $\sigma$). As the drop shrinks, its hydrophobic core becomes more pronounced, with increasing hydrophobicity values observed near the center. Reducing the volume of the drop means reducing its coefficients $\sigma_x$, $\sigma_y$, and $\sigma_z$ (Ref.2).[3–5]

The polypeptide chain folding simulation [3] proceeds as follows:



Published by Woodhead Publishing Limited, 2012

1. The starting structure is the early-stage intermediate, constructed on the basis of the contingency table.

   A. Two representations of the protein are prepared: an all-atom model, and a simplified model consisting of effective-atom coordinates (the arithmetic averages of the positions of all atoms belonging to each side chain).

   B. The molecule (in both representations) is immersed in a coordinate system in such a way that its geometric center coincides with the origin of the system, resulting in an average value of 0 in the Gaussian formula. The molecule is then rotated about each axis to ensure that the three orthogonal lines connecting the most distant elements (i.e., atoms) coincide with *X*, *Y*, and *Z* axes.

   C. The coefficients $\sigma_x$, $\sigma_y$, and $\sigma_z$ are calculated for the orientation obtained in step B. The distances between the origin of the coordinate system and the most distant atoms (along each axis) are increased by 9Å in each direction to account for the cutoff distance for hydrophobic interactions. As the calculations are performed separately for the positive and negative half-axes, the greater of the two values is used. This value is then divided by 3 and assumed to represent the coefficient $\sigma$ for the given axis.

   D. A cuboid grid is introduced, with an edge length equivalent to the greatest distance obtained in step C. By definition, this grid completely encapsulates the protein molecule.

   E. For each point of the grid established in step D, the value of the Gaussian function is calculated and then normalized by division by the aggregate sum of the values calculated for all points. For the *j*-th grid point, the resulting value is given by



Published by Woodhead Publishing Limited, 2012

$$\tilde{H}t_j = \frac{1}{\tilde{H}t_{sum}} \exp\left(\frac{-(x_j-\bar{x})^2}{2\sigma_x^2}\right)\exp\left(\frac{-(y_j-\bar{y})^2}{2\sigma_y^2}\right)\exp\left(\frac{-(z_j-\bar{z})^2}{2\sigma_z^2}\right)$$

F. The hydrophobic interaction of each grid point (its own hydrophobicity is equal to 0) with the adjacent effective atoms (representing side chains) is quantified using a cutoff distance of 9Å:

$$\tilde{H}o_j = \frac{1}{\tilde{H}o_{sum}} \times \sum_{i=1}^{N}(H_i^r)\begin{cases}\left[1-\frac{1}{2}\left(7\left(\frac{r_{ij}}{c}\right)^2 - 9\left(\frac{r_{ij}}{c}\right)^4 + 5\left(\frac{r_{ij}}{c}\right)^6 - \left(\frac{r_{ij}}{c}\right)^8\right)\right] & \text{for } r_{ij} \le c \\ 0 \text{ for } r_{ij} > c\end{cases}$$

Again, all values are normalized by division by the aggregate sum of the values assigned to all grid points. In this calculation, the "target" grid point's own hydrophobicity is assumed to be 0. Each grid point collects together the hydrophobic interactions of all side chains at distances below the cutoff.

G. Owing to the normalization of the two distributions, we can now meaningfully calculate the difference between the expected and observed hydrophobicities at each point:

$$\Delta\tilde{H} = \sum_{j=1}^{N}(\tilde{H}t_j - \tilde{H}o_j)^2$$

2. The optimization of the hydrophobicity distribution on a molecular level means minimizing the $\Delta\tilde{H}$ values. This process affects the entire molecule and is known as TM (total molecule) optimization. It is performed on the basis of the all-atom model. For each conformation,



Published by Woodhead Publishing Limited, 2012

the positions of the effective atoms (side chains) and the corresponding coefficients $\sigma$ are calculated.

3. Local optimization is performed, where the structure obtained in step 2 is optimized further from the point of view of intramolecular interactions. The optimization of the hydrophobicity distribution may have introduced unwanted local collisions. In order to eliminate these, each round of hydrophobicity optimization is followed by intramolecular optimization, affecting the electrostatic force (the first component in the following equation, where $q_i$ is the partial charge on the $i$-th atom, $\varepsilon_D$ is the dielectric constant, and $r_{ij}$ is the distance between the $i$-th and $j$-th atoms), the van der Waals force (the second component, where $\varepsilon_{ij}$ and $r_{0,ij}$ are polarizability parameters), and the torsion-potential force (the third component, where $K_\chi$ is the force constant, $\chi$ is the rotation angle, $n$ is the axial symmetry, and $\sigma$ is the phase):

$$V = \sum_{\text{all pairs}} \frac{q_i q_j}{\varepsilon_D r_{ij}} + \sum_{\text{all pairs}} \varepsilon_{ij} \left[ \left( \frac{r_{0,ij}}{r_{ij}} \right)^{12} - \left( \frac{r_{0,ij}}{r_{ij}} \right)^{6} \right] + \sum_{\text{torsionals}} K_\chi \left[ 1 + \cos(n\chi - \sigma) \right]$$

Optimization of the local interactions (LI) is performed for the all-atom representation, by calculating pairwise interactions between individual atoms. Note that the two optimization procedures (intramolecular and hydrophobic) remain in conflict with each other, as LI optimization may "upset" the structure returned by the TM optimization, and vice versa (because the positions of the effective atoms are taken into account in the TM optimization). Thus, each LI optimization step is followed by a TM optimization step where the volume of the bounding cube (the "drop") is progressively reduced.



Published by Woodhead Publishing Limited, 2012

The degree of this reduction can be assessed by comparing the ES intermediate with the crystalline protein form (depending on the length of its chain).

This phase of the folding simulation is finalized with a special, extended LI optimization step, which applies a more restrictive set of convergence criteria. The total number of iterations depends on the size of the starting structure (the volume of the bounding cube) and the length of the polypeptide chain.

## 7.2 Folding in the presence of a ligand

The conclusions drawn from the analysis of proteins that bind ligands or form protein–protein complexes point to the need to adjust further the "drop encapsulation" procedure, by introducing a factor that acts to deform the idealized structure of the hydrophobic core. Such deformations may take various forms – for example, they may be caused by the presence of ligands, characterized by their own hydrophobicity profiles. A highly hydrophobic ligand will compete with hydrophobic residues located near the center of the "drop," forcing the chain to alter its conformation; a ligand with dipolar hydrophobicity will try to align itself along the hydrophobicity gradient; and a poorly hydrophobic ligand will remain near the surface of the molecule. Each of these events is depicted in Figure 7.2.

It is also possible to introduce a deformation by placing a rigid protein molecule (a potential complexation target) in close proximity to the simulated chain. Similarly, it appears that the formation of highly specific catalytic sites in enzymes requires contact with a precursor molecule that must be present in the substrate at the folding stage.



Published by Woodhead Publishing Limited, 2012

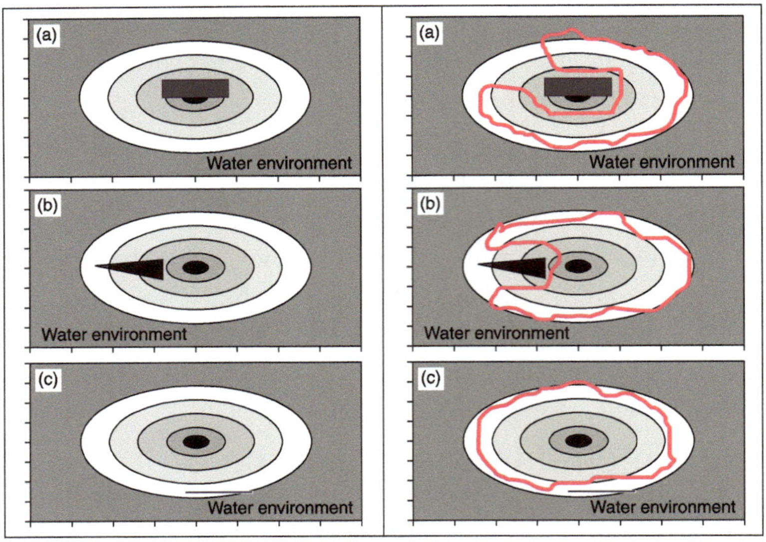


**Figure 7.2** Simulation of the folding process in the presence of a ligand. (a) Highly hydrophobic ligand; (b) ligand with variable (polarized) hydrophobicity; (c) hydrophilic ligand. The ellipses represent the isohydrophobic lines of the "fuzzy oil drop" model, and the shades of gray correspond to the hydrophobicity density (the darker the shade, the higher the hydrophobicity). The ligand's own hydrophobicity is reflected by the width of the strip by which it is marked: the thicker the strip, the higher the hydrophobicity in that particular part of the ligand. The right column shows the protein (red line) of fold influenced specifically by the presence of ligand.



Published by Woodhead Publishing Limited, 2012

## 7.3 Influence of external factors on polypeptide chain folding

As already mentioned, proteins whose structures are highly in accordance with the "fuzzy oil drop" model (including downhill and antifreeze proteins), while relatively uncommon, may serve as a convenient study group with which to validate our predictions. According to the explanation presented in the previous chapters, ideal accordance with the model would result in a molecule that was perfectly soluble (which may be a useful property) but unable to interact with other molecules present in its environment (which would clearly be detrimental). Analysis of the $\Delta\tilde{H}$ profiles and their reactions to the presence of external factors (Chapter 4) suggests that the $\Delta\tilde{H}$ profile reflects the deformations of the protein's hydrophobic core and may provide clues as to the nature of the causative factor. Thus, to model deformations of the protein's core accurately, one must take account of the presence of ligands (by "ligands," we mean any external factors that may affect the folding process so as to lead to the generation of a highly specific binding pocket). The "lock and key" abstraction, frequently invoked in describing the action of enzymes and their substrates (or inhibitors), applies here as well.

The "fuzzy oil drop" model must therefore be extended to include ligand molecules. Clearly, the ligand is characterized by its own hydrophobicity profile, which can be determined in a way analogous to the estimation of the hydrophobicity in the protein that is being analyzed. A highly hydrophobic ligand will compete with the protein's own peptides, trying to position itself at the center of the "drop." Dipolar ligands will align themselves along the hydrophobicity gradient, with their more hydrophobic parts buried deeper in the protein body. Finally, hydrophilic ligands may be attracted



Published by Woodhead Publishing Limited, 2012

to the surface of the protein molecule. In this way, the two components of the system (the polypeptide chain and the ligand) exert an influence on each other. The spectrum of possible outcomes is shown in Figure 7.2. The role of the ligand is to induce the polypeptide chain to fold in such a way as to accommodate the interacting molecule.

As mentioned before, the "ligand" may be a prosthetic group, a coenzyme, a substrate, an inhibitor, or an external protein molecule that forms a complex with the protein being analyzed. The definition can even be extended to include cellular membranes (which force the protein to assume a specific orientation or to embed itself in the membrane, facilitating its biological function). Simulations of ribonuclease and human hemoglobin, performed according to the model described here, were presented in Ref. 4 and 5, respectively. The validity and accuracy of this model have also been tested by conducting molecular dynamics simulations for the transmembrane protein 1U19 (bovine rhodopsin).[6] Results obtained using an all-atom model (acknowledging the presence of the membrane and the aqueous environment) indicate excellent (indeed, almost ideal) accordance with the results of studies in which the influence of the external force field is modeled as a 3D Gaussian function.[7]

Simulations of protein folding based on the "fuzzy oil drop" model have been performed for hemoglobin and ribonuclease both with and without ligands.[4,5] For hemoglobin, the positions of the histidine residues responsible for binding heme were correctly calculated, and for ribonuclease, the structure was characterized by the correct distribution of hydrophobicity in the presence of a ligand. A similar simulation for a transmembrane protein (specifically, TA0354_69_121 T0215, a hypothetical membrane protein in *T. acidophilum*, PDB code 1X9B) also produced satisfactory results although simulated without the presence of any external factor.[3]



Published by Woodhead Publishing Limited, 2012

In conclusion, the folding simulations conducted thus far seem to support the validity of our model while hinting at the need to extend its concepts further and perform tests on a wider variety of proteins. This will be a focus of future work for us.

## 7.4 References

1. Brylinski M., Konieczny L., Kononowicz A., Roterman I. (2008) Conservative secondary structure motifs already present in early-stage folding (*in silico*) as found in serpines family. *J. Theor. Biol.* 251(2), 275–85.
2. Brylinski M., Konieczny L., Roterman I. (2006) Fuzzy-oil-drop force field – a model to represent late-stage folding (*in silico*) of lysozyme. *J. Biomol. Struct. Dynam.* 23, 519–27.
3. Konieczny L., Brylinski M., Roterman I. (2006) Gauss-function-based model of hydrophobicity density in proteins. *In Silico Biol.* 6, 15–22.
4. Brylinski M., Konieczny L., Roterman I. (2006) Hydrophobic collapse in (*in silico*) protein folding. *Comput. Biol. Chem.* 30, 255–67.
5. Brylinski M., Konieczny L., Roterman I. (2007) Is the protein folding an aim-oriented process? Human haemoglobin as example. *Int. J. Bioinform. Res. Appl.* 3, 234–60.
6. Okada T., Sugihara M., Bondar A.N., Elstner M., Entel P., Buss V. (2004) The retinal conformation and its environment in rhodopsin in light of a new 2.2 Å crystal structure. *J. Mol. Biol.* 342, 571–83.
7. Zobnina V., Roterman I. (2009) Application of the fuzzy-oil-drop model to membrane protein simulation. *Protein Struct. Funct. Bioinform.* 77, 378–94.



Published by Woodhead Publishing Limited, 2012

## 7.5 Suggested reading

Banachewicz W., Religa T.L., Schaeffer R.D., Daggett V., Fersht A.R. (2011) Malleability of folding intermediates in the homeodomain superfamily. *Proc. Natl. Acad. Sci. USA* 108(14), 5596–601.

Brooks C.L. 3rd, Gruebele M., Onuchic J.N., Wolynes P.G. (1998) Chemical physics of protein folding. *Proc. Natl. Acad. Sci. USA* 95(19), 11037–8.

Bryan A.W. Jr., Starner-Kreinbrink J.L., Hosur R., Clark P.L., Berger B. (2011) Structure-based prediction reveals capping motifs that inhibit *β*-helix aggregation. *Proc. Natl. Acad. Sci. USA* 108(27), 11099–104.

Cheon S., Liang F. (2011) Folding small proteins via annealing stochastic approximation Monte Carlo. *Biosystems* 105(3), 243–9.

Christensen H., Pain R.H. (1991) Molten globule intermediates and protein folding. *Eur. Biophys. J.* 19(5), 221–9.

Creighton T.E. (1990) Protein folding. *Biochem. J.* 270, 1–16.

Fink A.L. (1999) Chaperone-mediated protein folding. *Physiol. Rev.* 79(2), 425–49.

Gaston D., Susko E., Roger A.J. (2011) A phylogenetic mixture model for the identification of functionally divergent protein residues. *Bioinformatics* 27(19), 2655–63.

Gazit E. (2002) The "Correctly Folded" state of proteins: is it a metastable state? *Angew. Chem. Int. Ed. Engl.* 41(2), 257–9.

Ghosh P.K., Li M.S., Bag B.C. (2011) Periodic force induced stabilization or destabilization of the denatured state of a protein. *J. Chem. Phys.* 135(11), 114101.

Gruebele M., Wolynes P.G. (1998) Satisfying turns in folding transitions. *Nat. Struct. Biol.* 5(8), 662–5.



Published by Woodhead Publishing Limited, 2012

Jäger M., Nguyen H., Crane J.C., Kelly J.W., Gruebele M. (2001) The folding mechanism of a beta-sheet: the WW domain. *J. Mol. Biol.* 311(2), 373–93.

Jonsson A.L., Daggett V. (2011) The effect of context on the folding of β-hairpins. *J. Struct. Biol.* 176(2), 143–50.

van der Kamp M.W., Daggett V. (2011) Molecular dynamics as an approach to study prion protein misfolding and the effect of pathogenic mutations. *Top. Curr. Chem.* 305, 169–97.

Karplus M. (2011) Behind the folding funnel diagram. *Nat. Chem. Biol.* 7(7), 401–4.

Lin M.M., Mohammed O.F., Jas G.S., Zewail A.H. (2011) Speed limit of protein folding evidenced in secondary structure dynamics. *Proc. Natl. Acad. Sci. USA* 108(40), 16622–7.

Lindorff-Larsen K., Piana S., Dror R.O., Shaw D.E. (2011) How fast-folding proteins fold. *Science* 334(6055), 517–20.

Morrone A., Giri R., Toofanny R.D., Travaglini-Allocatelli C., Brunori M., et al. (2011) GB1 is not a two-state folder: identification and characterization of an on-pathway intermediate. *Biophys J.* 101(8), 2053–60.

Motta A., Reches M., Pappalardo L., Andreotti G., Gazit E. (2005) The preferred conformation of the tripeptide Ala–Phe–Ala in water is an inverse gamma-turn: implications for protein folding and drug design. *Biochemistry* 44(43), 14170–8.

Naganathan A.N., Orozco M. (2011) The native ensemble and folding of a protein molten-globule: functional consequence of downhill folding. *J. Am. Chem. Soc.* 133(31), 12154–61.

Nishimura C., Dyson H.J., Wright P.E. (2006) Identification of native and non-native structure in kinetic folding intermediates of apomyoglobin. *J. Mol. Biol.* 355(1), 139–56.



Published by Woodhead Publishing Limited, 2012

Nivón L.G., Shakhnovich E.I. (2011) Thermodynamics and kinetics of the hairpin ribozyme from atomistic folding/unfolding simulations. *J. Mol. Biol.* 411(5), 1128–44.

Pawlowski K., Rychlewski L., Zhang B., Godzik A. (2001) Fold predictions for bacterial genomes. *J. Struct. Biol.* 134(2–3), 219–31.

Scott K.A., Daggett V. (2007) Folding mechanisms of proteins with high sequence identity but different folds. *Biochemistry* 46(6), 1545–56.

Scott K.A., Alonso D.O, Sato S., Fersht A.R., Daggett V. (2007) Conformational entropy of alanine versus glycine in protein denatured states. *Proc. Natl. Acad. Sci. USA* 104(8), 2661–6. Epub 2007 Feb 16.

Scott K.A., Randles L.G., Moran S.J., Daggett V., Clarke J. (2006) The folding pathway of spectrin R17 from experiment and simulation: using experimentally validated MD simulations to characterize states hinted at by experiment. *J. Mol. Biol.* 359(1), 159–73.

Scouras A.D., Daggett V. (2011) The Dynameomics rotamer library: amino acid side chain conformations and dynamics from comprehensive molecular dynamics simulations in water. *Protein Sci.* 20(2), 341–52.

Tian J., Garcia A.E. (2011) Simulation studies of protein folding/unfolding equilibrium under polar and nonpolar confinement. *J. Am. Chem. Soc.* 133(38), 15157–64.



Published by Woodhead Publishing Limited, 2012

8

# Misfolded proteins

*M. Król, L. Konieczny, K. Stąpor, Z. Wiśniowski, W. Ziajka, G. Szoniec, and I. Roterman-Konieczna*

**Abstract:** Structural analysis of properly folded proteins provides insight into their folding process and the origins of their biological properties; however, an equally rich trove of information presents itself in the form of misfolded proteins, particularly amyloidogenic ones. Such proteins exhibit structural modifications that induce the formation of large-scale fibrillary aggregates. Changes that promote amyloidogenesis also destroy the protein's intended biological activity. Relating the phenomenon of amyloid generation to the proposed folding model suggests the importance of F-type motifs, emerging as a result of structural forms introduced by external forces.

**Key words:** zone D, zone F, forced rotation, molecular dynamics simulations with external constraints, aggregation, $\alpha$-1-synuclein.

## 8.1 Introduction

The existence of amyloids is one of the fundamental problems related to protein folding. An amyloid may be created when



Published by Woodhead Publishing Limited, 2012

a protein changes its tertiary structure from the native structure to a pathological form. Amyloid clusters form complexes that consist of many individual monomers.[1] Unchecked growth of such complexes may lead to major disruption in the functioning of key biological machinery (particularly in respect to neuronal system including brain), causing disorders such as Alzheimer's disease and Creutzfeldt–Jakob disease.[2–4] Determining the origin of misfolded proteins remains a crucial challenge in studies of protein folding, as it provides evidence for major changes in the physiochemical properties of proteins (including their propensity to form aggregates) solely as a result of conformational shifts, which are occasionally (but not necessarily) triggered by mutations.[5]

It is often reported that $\beta$-like structures are particularly prone to amyloidogenesis. The reason for this is that they have been experimentally observed to form elongated fibrils, which, when subjected to complexation, tend to coil into spirals. Owing to their characteristic signature in radiograms, these spirals are sometimes referred to as cross-$\beta$ structures.[6–9]

Experimental studies also suggest that prolonged shaking of solvated proteins leads to aggregation and produces large amyloid clusters.[10] This process can be accelerated by the introduction of torsional forces acting on the polypeptide chains.[11] It is thought that any large domain positioned at the terminus of a $\beta$-twist promotes the generation of amyloids, as its mass presents an additional source of inertia that exacerbates torsion.

## 8.2 *In silico* experiment

This chapter presents an *in silico* experiment whose results, interpreted in the context of our model, may help explain the



Published by Woodhead Publishing Limited, 2012

phenomenon of amyloidogenesis. The *in silico* experiment involved the following steps:

1. A 1ns long molecular dynamics simulation was performed for a phenylalanine polypeptide starting from an extended initial conformation ($\Psi = \Phi = 180°$).
2. A representative structure (as obtained from step 1) was taken as the starting conformation for another 3 ns long molecular dynamics simulation, with an additional external constraint in the form of forced rotation.
3. Clustering of the freely rotating (unconstrained) region was performed, leading to the conclusion that zone F represents the most frequently occurring structural motif in the freely rotating fragment.
4. A final molecular dynamics simulation focusing on zone F (without any external constraints) was performed, producing a highly ordered $\beta$ structure with a large number of hydrogen bonds (which were also present in unordered areas). This simulation was performed on the C-terminal fragment of the highly amyloidogenic protein $\alpha$-1-synuclein.

A more detailed description of the steps of the experiment is given below:

1. An unrestrained molecular dynamics simulation of a polypeptide consisting of 30 phenylalanine residues produced a representative structure of the polypeptide, which was chosen for the next molecular dynamics simulation with a torsional constraint. (Details of the molecular dynamics simulation and the cluster analysis are given in the appendices to this chapter.)
2. The most representative structure obtained from the previous step provided the starting point for molecular



Published by Woodhead Publishing Limited, 2012

dynamics simulations (3ns in duration) with an additional external constraint in the form of a forced rotation. This is depicted schematically in Figure 8.1. The terminating regions (1–5 and 26–30) were kept rigid. The central region (13–18) was subjected to forced rotation. The intermediate regions (6–12 and 19–25) remained free of any imposed constraints. The goal of the analysis was to determine the conformation of the regions denoted by W (free) in the figure as a result of interaction with the regions denoted by S (rigid/fixed) and K (forced rotation).

3. The values of $\Psi$ and $\Phi$ in the freely rotating regions were calculated for structural snapshots along the trajectory and were subjected to hierarchical clustering, where the measure $D_{avg}$ was used to identify clusters and elucidate the most frequently occurring values of the angles $\Psi$ and $\Phi$, which appeared to be (–92, +142), (–91, +136), (–85,

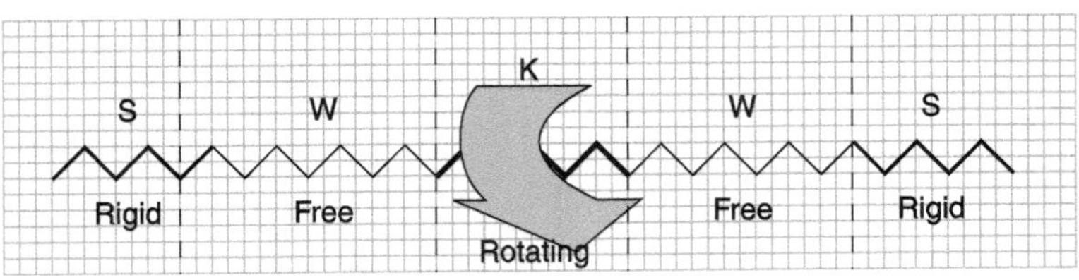


**Figure 8.1** Schematic depiction of the polyphenylalanine polypeptide that was the subject of the study presented here. The regions 1–5 and 26–30, labeled "S," were assumed to be rigid (intermediate-thickness lines); the central region 13–18, labeled "K," was subjected to torsional forces (thick line); and the intermediate regions 6–12 and 19–25 could adapt freely to external forces (thin lines)



Published by Woodhead Publishing Limited, 2012

+145), and (−139, +135). The relative frequencies of the angles $\Phi$, $\Psi$ representing particular zones distinguished on the Ramachandran map (see Figure 3.2) are presented in Figure 8.2.

Analysis of the likelihood of encountering various structural motifs in specific regions (Figure 8.2) indicated that the torsion zones were dominated by F-type structures, in addition to a lesser number of right-handed (C) and left-handed (G) helical motifs. It should be noted that class F, despite being adjacent to zone E, exhibits peculiar morphological features (Figure 8.3). Figure 8.3 illustrates the various spatial arrangements of atoms capable of forming hydrogen bonds. In an F-type structure, the chain becomes twisted and thus differs from a standard $\beta$-strand. The planar conformation of C=O (red) and N–H ( blue) bonds in the bottom image ($\beta$ structure) facilitates the formation of a flat sheet, whereas in the F-type structure (top), the resulting dimeric chain becomes a twisted spiral with a fixed, relatively low radius of curvature (particularly when compared with the very large radius of curvature for $\beta$ structures).

4. The most interesting example (in the context of the model presented in this book) is the observation of a high frequency of F-zone structural motifs. It should be noted that, as previously mentioned, these structures were obtained in a forced molecular dynamics simulation, which is in agreement with the results of experiments leading to amyloid formation *in vitro*.[10,11] To link this observation with the next observation, in which we recognized $\alpha$-1-synuclein as an amyloidogenic protein, a simulation was performed for the loose C-terminal region (which consists of 47 amino



Published by Woodhead Publishing Limited, 2012

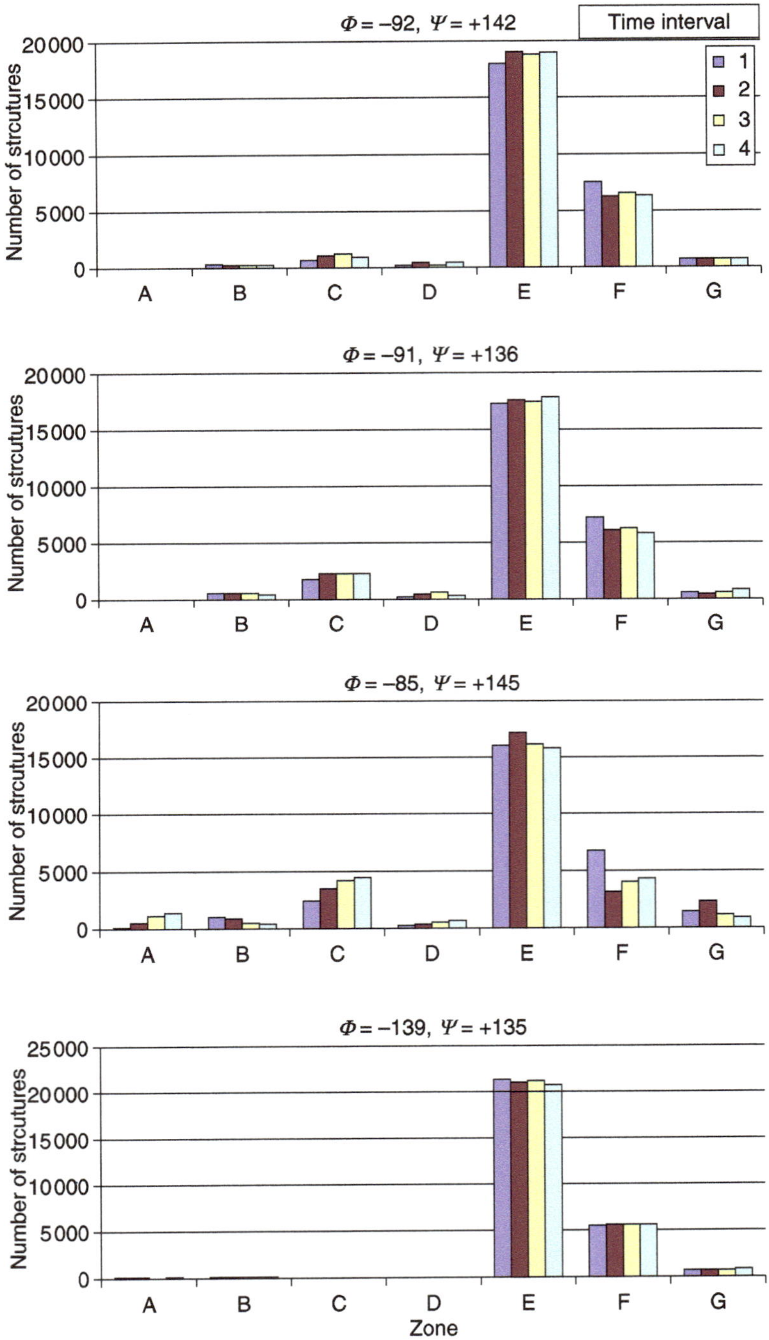


**Figure 8.2** Caption opposite



Published by Woodhead Publishing Limited, 2012

acids) of this protein.[12,14] Angles $\Psi$ and $\Phi$ of −60° and 135°, respectively, were assumed in the starting structure, in line with the observed probability of occurrence of specific values of $\Psi$ and $\Phi$ in zone F (Chapter 3). The goal of the molecular dynamics simulation was to test the protein's capacity for polymerization: the calculations were expected to produce a dimeric structure capable of generating multichain aggregates.

This part of the experiment involved two chains of 47 amino acids each (the angles $\Psi$ and $\Phi$ for all residues in this polypeptide were taken as −60° and 135°, respectively), consistent with the unstructured C-terminal region of $\alpha$-1-synuclein (1XQ8) (the helical N-terminal region was omitted). Parallel and antiparallel mutual orientations of the polypeptides (referred to as the conformations PP and AP, respectively) were considered. The source chain was first oriented along the *Y* axis and then paired with a complementary chain, in both conformations, with displacements in the *XY* and *YZ* planes to ensure varied interchain contacts. This phase resulted in four structures, labeled PP*XY*, PP*YZ*, AP*XY*,

**Figure 8.2** Frequency of occurrence of each structural motif (quantified by the number of conformations identified for each letter of the structural alphabet) in the postdynamics structures produced by molecular dynamics simulations in the presence of torsional forces (illustrated in Figure 8.1). Only the intermediate (W) regions were considered. The $\Psi$ and $\Phi$ values listed in the headers indicate the initial conditions for each simulation



Published by Woodhead Publishing Limited, 2012

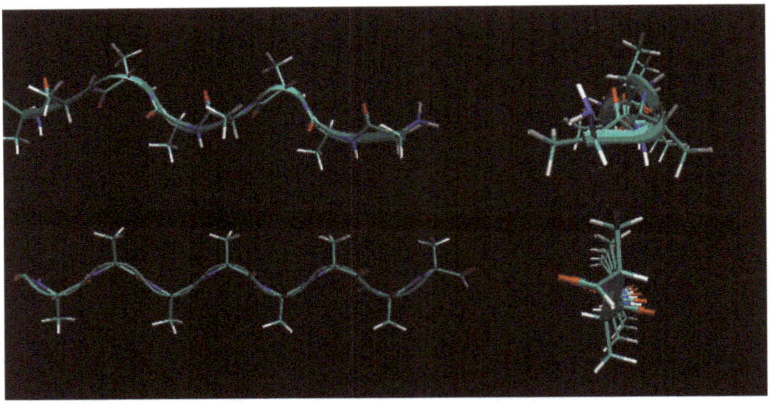

**Figure 8.3** **Sample polypeptide fragments constructed for $\Psi$ and $\Phi$ values of −60, 135 (F-type structure) and −139, 135 (classic antiparallel $\beta$-sheet). Left: different arrangements of atoms capable of forming hydrogen bonds; right: 3D view along the chain axis. The program VMD was used for visualization**

and APYX. Calculations were performed using the CHARMM tool with the EEF1 implicit solvation model over a period of 2 ns. Radical structural changes were observed during the first nanosecond. The angles $\Psi$ and $\Phi$ were calculated for the final 100 ps of the simulation, and the final results were represented in the PDB format.

The simulation results are depicted in Figure 8.4. The study indicates that $\beta$ structures are generated quite readily. Of note is the viability of both parallel and antiparallel $\beta$-structural forms (regardless of the starting structure). Analysis of the lattice of hydrogen bonds suggests that these bonds play a significant role in stabilizing the aggregate, particularly in fragments that do not exhibit $\beta$-type ordering (Figure 8.5) – in fact,



Published by Woodhead Publishing Limited, 2012

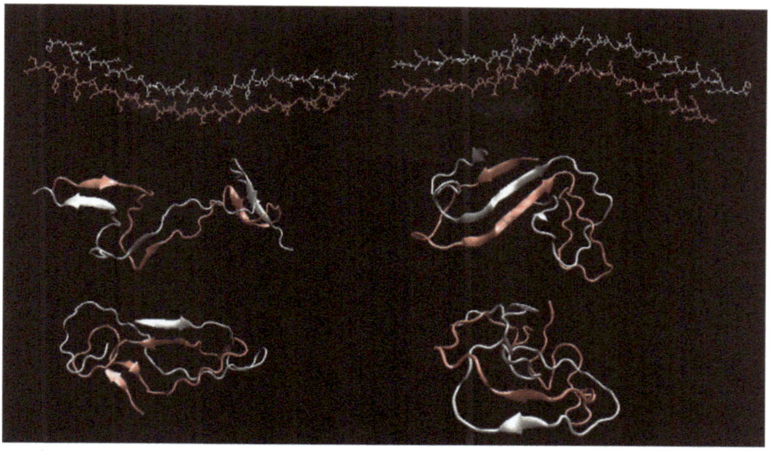

**Figure 8.4** Postdynamics structural forms of the C-terminal region (47 amino acids) of $\alpha$-1-synuclein. Antiparallel (left) and parallel (right) conformations of the two-chain complex are shown. Top row, starting structures; central row, displacement in the *YZ* plane; bottom row, displacement in the *XY* plane. The program VMD was used

hydrogen bonds were found along the entire polypeptide chain.

The presence of numerous $\beta$-structural regions and the abundance of hydrogen bonds suggests that such highly ordered structures may propagate easily, forming much larger aggregates than those considered in our simulation. Although the starting structures were of type F, none of the postdynamics structures were found to belong to this zone (Figure 8.6).

Summing up the analysis presented here, it appears that the stability of amyloid aggregates hinges upon the properties of



Published by Woodhead Publishing Limited, 2012

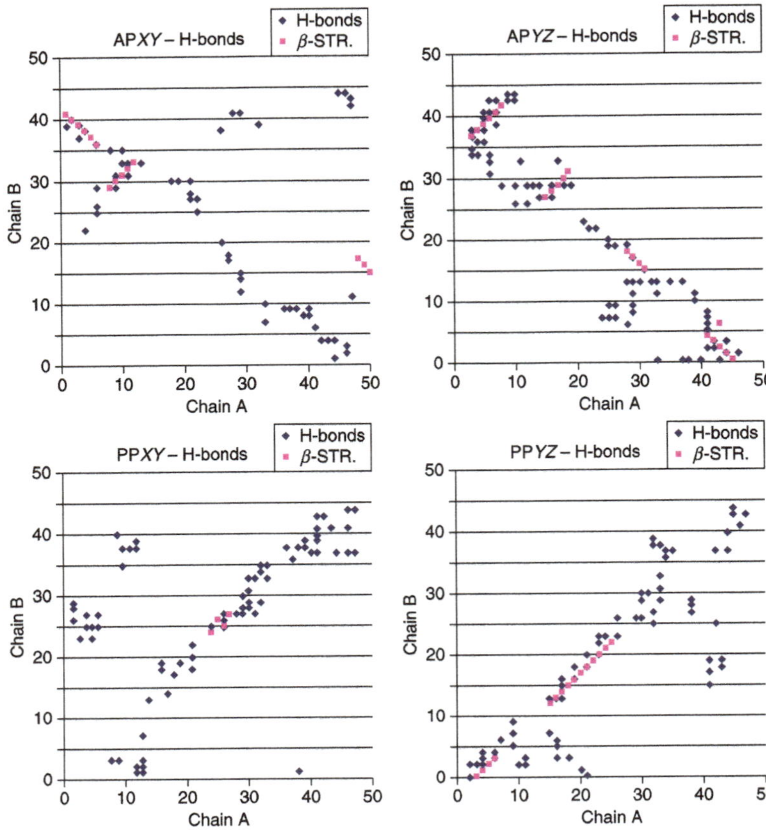


**Figure 8.5** **Hydrogen bond lattice in the postdynamics structure for each of the four conformations presented in Figure 8.4. Each dot represents a pair of amino acids that form a hydrogen bond. Pink dots correspond to amino acids interacting through hydrogen bonds within *β*-type structural motifs**

coherent *β*-type fragments, while the remaining (poorly ordered) protein regions are instead stabilized by a network of hydrogen bonds. Despite their apparently chaotic nature, such regions can propagate if interspersed with short *β*



Published by Woodhead Publishing Limited, 2012

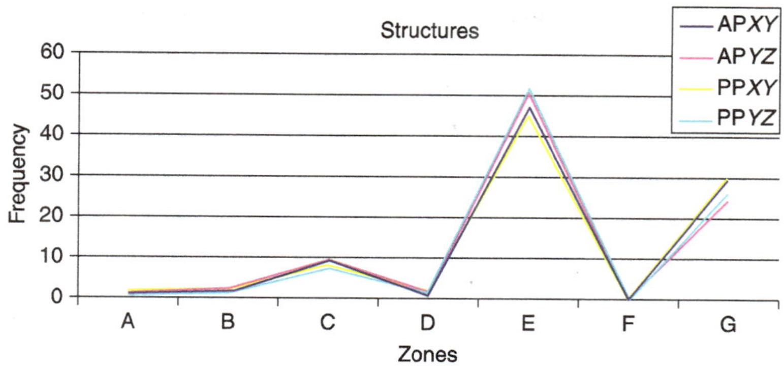


**Figure 8.6** Quantitative comparison of the frequency of occurrence of F-type structural motifs in the postdynamics conformations for a dimeric C-terminal region of $\alpha$-1-synuclein

fragments, leading perhaps to the formation of amyloid plaques of considerable size.

## 8.3 Conclusions

By referring to the mechanisms described in this chapter and the structural properties of proteins, which can be obtained from databases, we can try to describe the general characteristics of F-type structures as follows:

1. F-type structures exist most frequently (though not always) as single amino acids. The longest F fragment found so far consists of three adjacent peptides.
2. Single F-type amino acids usually occur amid class E sequences (the classic $\beta$ conformation), although there are also $\beta$-structural fragments that consist entirely of E-type motifs, without any F-type residues.



Published by Woodhead Publishing Limited, 2012

3. Another conspicuous property of F-type structures is that they often appear in terminal regions of long $\beta$-sheets (these are akin to D-type structures, which can frequently be observed near the termini of helical regions – see Figure 3.4).

Several sample proteins (randomly chosen from polypeptide chains that exhibit high concentrations of F-type structural motifs) are depicted in Figure 8.7. The distribution of the angles $\Psi$ and $\Phi$ in 3PO3 suggests a conformational continuum between zones E and F, whereas in 3J00, 3N7U, and 3AV5 there are distinct sets of ($\Phi$, $\Psi$) values that belong to zone F (in the case of 3AV5, this set is even more populous than the corresponding E set, which represents classic $\beta$-type structures). This observation justifies treating zone F as materially different from zone E. The proteins represented in Figure 8.7 are as follows:

- 3J00 (98 amino acids + 1630 bases). A ribosome/ribosomal protein.This is part of a ribosome–SecYE complex in the membrane environment with 16s ribosomal RNA.[15]
- 3N7U (4212 amino acids). An oxidoreductase. This is an NAD-dependent formate dehydrogenase from the higher plant *Arabidopsis thaliana* in complex with NAD and azide.[16]
- 3AV5 (1139 amino acids). A transferase. This is described in PDBsum as mouse DNA methyltransferase 1 with S-adenosyl-L-homocysteine DNA (cytosine-5)-methyltransferase.[17]
- 3PO3 (1426 amino acids + 25 bases). A transferase. This is a DNA/RNA arrested RNA polymerase II reactivation intermediate.[18]



Published by Woodhead Publishing Limited, 2012

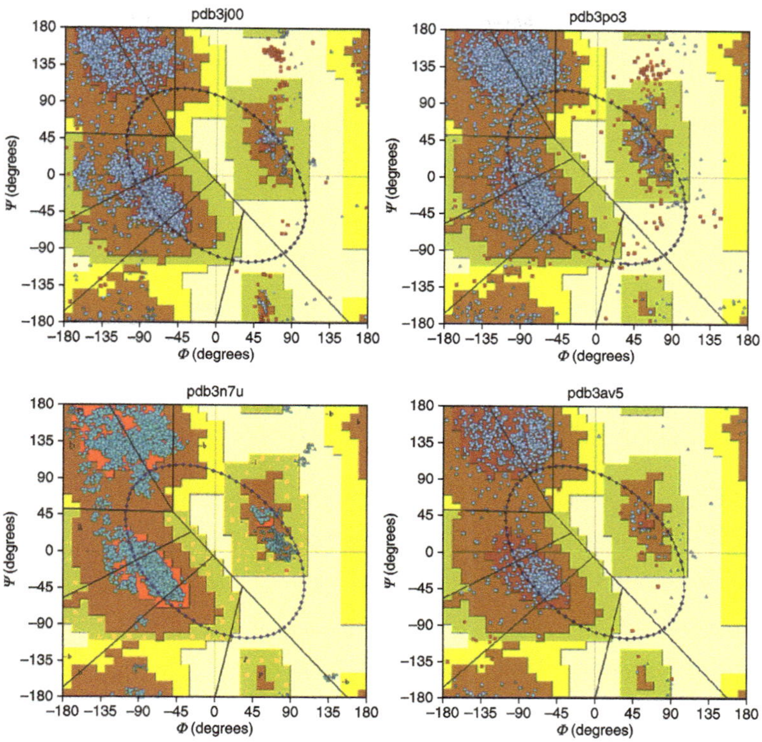


**Figure 8.7** Four Ramachandran maps plotted for proteins with various representations of F-type structural motifs. In the case of 3J00, 3N7U, and 3AV5, the plot shows distinct sets of $\Psi$ and $\Phi$ values that fall within zone F, whereas in the case of 3P03 the E and F populations form a single, common set. The initial plots were obtained from PDBsum (*http://www.ebi.ac.uk/pdbsum/*). The elliptical paths and zone limits are in agreement with the methodology presented in Chapter 3. The plots illustrate the frequency of occurrence of individual letters of the structural alphabet in proteins formed in the presence of torsional forces (see Figure 3.4)



Published by Woodhead Publishing Limited, 2012

Analysis of the distributions of the angles $\Psi$ and $\Phi$ (particularly within zone F) suggests that such motifs are most likely to be formed under the influence of external forces (by complexation with nucleic acids, binding numerous ligands, or polymerization) and that mitigating such factors would instead yield a $\beta$-rich structure. This hypothesis requires further study, which will be a subject of future work.

## 8.4 Appendix 1: details of the molecular dynamics simulation

The starting structure of the polyphenylalanine polypeptide was built in the extended conformation ($\Phi = \Psi = 180°$) and consisted of 30 PHE residues. Prior to any further calculations, the polypeptide molecule was energy-minimized in a GB/SA implicit solvent model to relieve any bad atom contacts. Subsequently, the molecule was immersed in a rectangular box of TIP3P water. The minimum distance between any atom of the molecule and the box was set to 10 Å. $Na^+$ and $Cl^-$ ions were added to obtain a salt concentration of roughly 0.1M. The resulting system was energy-minimized and equilibrated by a 100 ps NPT molecular dynamics simulation. During the minimization and equilibration, the protein backbone atoms were restrained to avoid large distortions of the polypeptide molecule. After equilibration of the system, a 1ns long unrestrained simulation was performed to obtain a sample conformation of the polypeptide in solution corresponding to the $\beta$ secondary structure. The sample conformation was taken as the starting structure for the restrained simulations. Rotational restraints were applied to the central six residues (residues 13–18) of the polypeptide to introduce rotational strain in the molecule. The axis of rotation was roughly aligned with the extended backbone of



Published by Woodhead Publishing Limited, 2012

the polypeptide, and the rotational velocity was set to one full rotation (360°) per nanosecond of the simulation. To prevent the whole molecule from rotating as a rigid body, the C- and N-terminal residues (residues 1–5 and 25–30) were kept fixed, while residues 6–12 and 19–25 were unrestrained and free to rotate with the restrained residues 13–18. In this way, we observed a buildup of rotational strain in the unrestrained fragments of the molecule. In order to sample different strained conformations, we ran two independent 2 ns long simulations, starting with different initial velocities. Snapshots were collected every picosecond.

All explicit solvent simulations were performed using the program NAMD[19] with the CHARMM22 all-atom force field.[20] The pressure and temperature were kept at 1 atm and 298 K, respectively, resulting in an NPT simulation system. All bond lengths between hydrogens and heavy atoms were fixed using the SHAKE algorithm, which allowed a time step of 2 fs.

Values of the dihedral angles $\Phi$ and $\Psi$ were calculated for the unrestrained region (residues 6–12 and 19–24) for all snapshots taken during the two molecular dynamics simulations. The values were clustered, and the center of each cluster was taken as a representative value of the angles $\Phi$ and $\Psi$ of either the restrained or the relaxed structure.

For each cluster center, an alanine polypeptide 20 amino acids long was built, with all dihedral angles $\Phi$ and $\Psi$ equal to the cluster center values. In total, eight different polypeptides were built. Each polypeptide was duplicated, and two orientations of each pair (i.e., the polypeptide and a duplicate of it) were manually constructed: a parallel orientation, where the N- and C-termini of the two chains were close to each other, and an antiparallel orientation, where the N-terminus of one chain was close to the C-terminus of the other chain. The average distance between the two chains was fixed at 5 Å. For each orientation, a 3 ns long molecular dynamics simulation was performed. As the aim of these simulations was to investigate whether the chains were able to form stable $\beta$-sheet-like structures, we decided to use the EEF implicit solvent model [21] to accelerate the possible formation of interchain interactions. During the simulations, all bond lengths between hydrogens and heavy atoms were fixed using the SHAKE algorithm, which allowed a time step of 2 fs. The EEF simulations and all trajectory analyses were performed using the CHARMM molecular modeling package.[22] Snapshots were collected every picosecond.

Values of the angles $\Phi$ and $\Psi$ were calculated and interchain hydrogen bonds were identified for each snapshot of the resulting trajectories. The hydrogen bonds were analyzed to obtain the fraction of the simulation time at which each hydrogen bond was present, its average lifetime, and how many times the bond was formed during the simulation.

## 8.5 Appendix 2: details of the cluster analysis

A (hard) *clustering* into $c$ clusters ($c$-*clustering*) of a $n$-element set,

$$u_n = \{x_i = (x_{i1}, \ldots, x_{ip}) \mid i = 1, \ldots, n\}$$

where each $x_i$ is a $p$-dimensional vector of measured or computed object features, is defined as a set of $c$ nonempty and disjoint groups $G_i$ (known as *clusters*),

$$Cls^c(n) = \{G_1, \ldots, G_c \mid G_i \subseteq u_n, i = 1, \ldots, c\}$$

composed of vectors from the set $u_n$. In addition, the vectors contained in a cluster $G_i$ are "more similar" to each other



Published by Woodhead Publishing Limited, 2012

and "less similar" to the vectors in the other clusters. The most common *similarity measure d* between two vectors $x_i$, $y_j$ is their Euclidean distance

$$d(x_i, y_j) = \sqrt{\sum_{r=1}^{p} (x_{ir} - y_{jr})^2}$$

The similarity measure between a vector $x$ and a cluster $G$ can be defined as the maximum, minimum, or average distance in the set of distances between all pairs of vectors of $G$:

$$D_{\max/\min}(x, G) = \max/\min_{y \in G} \{d(x, y)\}$$

$$D_{avg}(x, G) = \frac{1}{n_G} \sum_{y \in G} d(x, y)$$

where $n_G$ is the cardinality of $G$, and $d(x, y)$ is a measure of the similarity between the two vectors. A similarity measure between two clusters $G_i$ and $G_j$ can be defined in a similar way:

$$D_{\max/\min}(G_i, G_j) = \max/\min_{x \in G_i, y \in G_j} \{d(x, y)\},$$

$$D_{avg}(G_i, G_j) = \frac{1}{n_{G_i} n_{G_j}} \sum_{x \in G_i} \sum_{y \in G_j} d(x, y).$$

Hierarchical clustering algorithms produce a hierarchy of nested clusterings known as a *dendrogram*, which is a binary tree. In the most common scheme, the initial clustering $Cls^n{}_0$ $(n)$ consists of $n$ clusters, each containing a single vector from a set $u_n$, and creates level 0 of the hierarchy (i.e., the dendrogram). In the first step, the clustering $Cls^{n-1}{}_1$ $(n)$, with $(n - 1)$ clusters, is produced as a result of joining the two most similar clusters in the previous clustering $Cls^n{}_0$ $(n)$, which gives level 1 of the growing hierarchy. This procedure continues until the final clustering, $Cls^1{}_{n-1}$ $(n)$, is obtained, which contains one cluster, a single set $u_n$. This scheme can be stated in the following algorithm.

1. $t = 0$. /* hierarchy level number*/
2. Create the initial clustering

   $$Cls_0^n(n) = \{G_i = \{x_i\} \mid i = 1, \ldots, n\}.$$

3. Create the initial *similarity matrix P(t)*, in which the element $(i, j)$ is equal to the distance between the joined clusters:

   $$P(i, j)(t) = D(G_i, G_j),\ i, j = 1, \ldots, n.$$

4. Repeat steps (a)–(e) until all vectors lie in a single cluster:

   (a) among all pairs of clusters $(G_r, G_s)$ in the clustering $t$ find the one, say $(G_i, G_j)$, such that

   $$D(G_i, G_j) = \min_{r,s} D(G_r, G_s);$$

   (b) $t = t + 1$;
   (c) create a new cluster $G_q = G_i \cup G_j$ from the most similar clusters $G_i$ and $G_j$;
   (d) create the new clustering

   $$Cls_t^{n-t}(n) = \{Cls_{t-1}^{n-t+1}(n) - \{G_i, G_j\}\} \cup \{G_q\};$$

   (e) update the similarity matrix $P(t)$ based on the previous $P(t-1)$ in the following way:

   (e1) remove the two rows and the columns that correspond to the joined clusters from the matrix $P(t-1)$;
   (e2) add the new row and the column for the newly created cluster. These should contain the computed distances between the new cluster



Published by Woodhead Publishing Limited, 2012

and all those remaining from the previous step ($t$ – 1), i.e., those which have not been changed.

See Ref. 23 and 24 for more details.

## 8.6 References


1. Dobson C.M. (2003) Protein folding and misfolding. *Nature* 426(6968), 884–90.
2. Berchtold N.C., Cotman C.W. (1998). Evolution in the conceptualization of dementia and Alzheimer's disease: Greco-Roman period to the 1960s. *Neurobiol. Aging* 19(3), 173–89.
3. Brookmeyer R., Johnson E., Ziegler-Graham K., Arrighi H.M. (2007). Forecasting the global burden of Alzheimer's disease. *Alzheimer's Dement.* 3(3), 186–91.
4. Jones M., Peden A.H., Prowse C.V. (2007). In vitro amplification and detection of variant Creutzfeldt–Jakob disease PrPSc. *J. Pathol.* 213(1), 21–6.
5. Prusiner S.B. (1998). Prions. *Proc. Natl. Acad. Sci. USA* 95(23), 13363–83.
6. Kirschner D.A., Abraham C., Selkoe D.J. (1986) X-ray diffraction from intraneuronal paired helical filaments and extraneuronal amyloid fibers in Alzheimer disease indicates cross-$\beta$ conformation. *Proc. Natl. Acad. Sci. USA* 83, 503–7.
7. Jahn T.R., Makin O.S., Morris K.L., Marshall K.E., Tian P., et al. (2009) The common architecture of cross-beta amyloid. *J. Mol. Biol.* 395(4), 717–27.
8. Nelson R., Sawaya M.R., Balbirnie M., Madsen A.Ø., Riekel C., et al. (2005) Structure of the cross-beta spine of amyloid-like fibrils. *Nature* 435(7043), 773–8.




Published by Woodhead Publishing Limited, 2012

9. Trovato A., Chiti F., Maritan A., Seno F. (2006) Insight into the structure of amyloid fibrils from the analysis of globular proteins. *PLoS Comput. Biol.* 2(12), e170.
10. Nilsson M.R. (2004) Techniques to study amyloid fibril formation in vitro. *Methods* 34, 151–60.
11. Schlunegger M.P., Bennet M.J., Eisenberg D. (1997) Oligomer formation by 3D domain swapping: a model for protein assembly and misassembly. In Richards F.M., Eisenberg D.S., and Kim P.S. (eds), *Protein Misassembly*, pp. 61–122, Advances in Protein Chemistry, Vol. 50. Amsterdam: Elsevier.
12. Apetri M.M., Maiti N.C., Zagorski M.G., Carem P.R., Anderson V.E. (2006) Secondary structure of $\alpha$-synuclein oligomers: characterization by Raman and atomic force microscopy. *J. Mol. Biol.* 355, 63–71.
13. Kaylor J., Bodner N., Edridge S., Yamin G., Hong D.P., Fink A.L. (2005) Characterization of oligomeric intermediates in alpha-synuclein fibrillation: FRET studies of Y125W/Y133F/Y136F alpha-synuclein. *J. Mol. Biol.* 353(2), 357–72.
14. Yamin G., Munishkina L.A., Karymov M.A., Lyubchenko Y.L., Uversky V.N., Fink A.L. (2005) Forcing nonamyloidogenic beta-synuclein to fibrillate. *Biochemistry* 44(25), 9096–107.
15. Frauenfeld J., Gumbart J., Van Der Sluis E.O., Funes S., Gartmann M., et al. Cryo-EM structure of the ribosome–SecYE complex in the membrane environment. *Nat. Struct. Mol. Biol.* 18, 614–21.
16. Shabalin I.G., Polyakov K.M., Skirgello O.E., Tishkov V.I., Popov V.O. Structures of the apo and holo forms of NAD-dependent formate dehydrogenase from the higher-plant *Arabidopsis thaliana*. Available from: *http://www.rcsb.org/pdb/explore/explore.do?structureId=3n7u* [accessed 7 May 2012].




Published by Woodhead Publishing Limited, 2012

17. Takeshita K., Suetake I., Yamashita E., Suga M., Narita H., et al. (2011) Structural insight into maintenance methylation by mouse DNA methyltransferase 1 (Dnmt1). *Proc. Natl. Acad. Sci. USA*. Available from: *http://www.rcsb.org/pdb/explore/explore.do?structureId=3av5* [accessed 7 May 2012].
18. Cheung A.C.M., Cramer P. (2011). Structural basis of RNA polymerase II backtracking, arrest and reactivation. *Nature* 471, 249–53.
19. Phillips J.C., Braun R., Wang W., Gumbart J., Tajkhorshid E., et al. (2005) Scalable molecular dynamics with NAMD. *J. Comput. Chem.* 26, 1781–1802.
20. MacKerell A.D.J., Bashford D., Bellot M., Dunbrack R.L.J., Evenseck J.D., et al. (1998) All-atom empirical potential for molecular modeling and dynamics studies of proteins. *J. Phys. Chem. B* 102, 3586–616.
21. Lazaridis T., Karplus M. (1999) Effective energy function for proteins in solution. *Proteins* 35, 133–52.
22. Brooks B.R., Brooks C.L. III, Mackerell A.D. Jr., Nilsson R., Petrella R.J., et al. (2009) CHARMM: the biomolecular simulation program. *J. Comput. Chem.* 30(10), 1545–614.
23. Jain A.K., Dubes R.C. (1988) *Algorithms for Clustering Data*. Englewood Cliffs, NJ: Prentice Hall.
24. Stąpor K. (2011) *Classification Algorithms in Computer Vision* [in Polish]. Warsaw: PWN Scientific Publishing.

## 8.7 Suggested reading

Andersen C.A., Palmer A.G., Brunak S., Rost B. (2002) Continuum secondary structure captures protein flexibility. *Structure* 10(2), 175–84.



Published by Woodhead Publishing Limited, 2012

Armen R.S., Daggett V. (2005) Characterization of two distinct beta2-microglobulin unfolding intermediates that may lead to amyloid fibrils of different morphology. *Biochemistry* 44(49), 16098–107.

Bartels T., Choi J.G., Selkoe D.J. (2011) $\alpha$-Synuclein occurs physiologically as a helically folded tetramer that resists aggregation. *Nature* 477, 107–10.

Capriotti E., Altman R.B. (2011) Improving the prediction of disease-related variants using protein three-dimensional structure. *BMC Bioinform.* Suppl. 4, S3.

Chun Wu, Zhixiang Wang, Hongxing Lei, Wei Zhang, Yong Duan (2007) Dual binding modes of Congo red to amyloid protofibril surface observed in molecular dynamics simulations. *J. Am. Chem. Soc.* 129, 1225–32.

Daidone I., Di Nola A., Smith J.C. (2011) Molecular origin of Gerstmann–Sträussler–Scheinker syndrome: insight from computer simulation of an amyloidogenic prion peptide. *Biophys J.* 100(12), 3000–7.

Gibson T.J., Murphy R.M. (2005) Design of peptidyl compounds that affect beta-amyloid aggregation: importance of surface tension and context. *Biochemistry* 44(24), 8898–907.

Haspel N., Zanuy D., Ma B., Wolfson H., Nussinov R. (2005) A comparative study of amyloid fibril formation by residues 15–19 of the human calcitonin hormone: a single beta-sheet model with a small hydrophobic core. *J. Mol. Biol.* 345(5), 1213–27.

Hohenester E., Hutchinson W.L., Pepys M.B., Wood S.P. (1997) Crystal structure of a decameric complex of human serum amyloid P component with bound dAMP. *J. Mol. Biol.* 269(4), 570–8.

Lashuel H.A., Wurth C., Woo L., Kelly J.W. (1999) The most pathogenic transthyretin variant, L55P, forms amyloid




Published by Woodhead Publishing Limited, 2012

fibrils under acidic conditions and protofilaments under physiological conditions. *Biochemistry* 38, 13560–73.

Lattman E.E., Fiebig K.M., Dill K.A. (1994) Modeling compact denatured states of proteins. *Biochemistry* 33(20), 6158–66.

Li L., Darden T.A., Bartolotti L., Kominos D., Pedersen L.G. (1999) An atomic model for the pleated beta-sheet structure of Abeta amyloid protofilaments. *Biophys. J.* 76(6), 2871–8.

Lipfert J., Franklin J., Wu F., Doniach S. (2005) Protein misfolding and amyloid formation for the peptide GNNQQNY from yeast prion protein Sup35: simulation by reaction path annealing. *J. Mol. Biol.* 349(3), 648–58.

López de la Paz M., de Mori G.M., Serrano L., Colombo G. (2005) Sequence dependence of amyloid fibril formation: insights from molecular dynamics simulations. *J. Mol. Biol.* 349(3), 583–96.

Ma B., Nussinov R. (2006) Simulations as analytical tools to understand protein aggregation and predict amyloid conformation. *Curr. Opin. Chem. Biol.* 10(5), 445–52.

Małolepsza E.B. (2008) Modeling of protein misfolding in disease. *Methods Mol. Biol.* 443, 297–330.

Massi F., Straub J.E. (2001) Energy landscape theory for Alzheimer's amyloid beta-peptide fibril elongation. *Proteins* 42(2), 217–29.

Nelson R., Sawaya M.R., Balbirnie M., Madsen A.Ø., Riekel C., et al. (2005) Structure of the cross-beta spine of amyloid-like fibrils. *Nature* 435, 773–8.

Oh S.Y., Ellenstein A., Chen C.D., Hinman J.D., Berg E.A., et al. (2005) Amyloid precursor protein interacts with notch receptors. *J. Neurosci. Res.* 82(1), 32–42.

Sebastião M.P., Merlini G., Saraiva M.J., Damas A.M. (2000) The molecular interaction of 4′-iodo-4′-deoxydoxorubicin with Leu-55Pro transthyretin



Published by Woodhead Publishing Limited, 2012

"amyloid-like" oligomer leading to disaggregation. *Biochem. J.* 351(Pt. 1), 273–9.

Serpell L.C., Blade C.C., Fraser P.E. (2000) Molecular structure of a fibrillar Alzheimer's A$\beta$ fragment. *Biochemistry* 39, 13269–75.

Sipe J.D. (1992) Amyloidosis. *Annu. Rev. Biochem.* 61, 947–75.

Tjernberg L.O., Callaway D.J.E., Tjernberg A., Halne S., Lilliehöök C., et al. (1999) A molecular model of Alzheimer amyloid $\beta$-peptide fibryl formation. *J. Biol. Chem.* 274, 12619–25.

Zhou H.X. (2004) Polymer models of protein stability, folding, and interactions. *Biochemistry* 43(8), 2141–54.



Published by Woodhead Publishing Limited, 2012

9

# A short description of other selected *ab initio* methods for protein structure prediction

*I. Roterman-Konieczna*

**Abstract:** This chapter discusses three models representing the *ab initio* approach: UNRES, the lattice model, and the model implemented by the ROSETTA package. Given the fact that some sources treat the tertiary conformation that corresponds to the protein's global free-energy minimum as equivalent to its native form, selected algorithms geared towards identifying such global minima will also be discussed.

**Key words:** *ab initio* model, united atoms, lattice model, structural motifs, field deformations, CASP, CAPRI, global minimum.

## 9.1 Introduction

In general terms, protein structure prediction is based on two competing paradigms. The Darwinian approach (sometimes referred to as "homology search" or "comparative modeling") assumes that proteins that represent successive stages of evolution must retain some common structural



Published by Woodhead Publishing Limited, 2012

properties in order to consistently perform their biological function. Thus, identifying evolutionary sequences of protein homologues is viewed as the key to predicting the structures of target proteins. By aggregating structural fragments that exhibit a high degree of sequential similarity, one can produce suitable starting structures for energy optimization procedures.

The Darwinian view competes with the so-called Boltzmann approach, which claims that structural modeling should be based solely on a theoretical model that, together with an appropriate energy minimization criterion, is expected to produce the correct structure for the polypeptide under consideration. Boltzmann-based algorithms are sometimes called "*ab initio*" or "new fold" models, as they attempt to predict the structural properties of proteins purely by means of deterministic simulations, with no reliance on protein databases. These names in quote marks specifically exclude any notion of heuristic analysis of existing protein structures. Instead, they posit that folding should be entirely dependent on the sequential arrangement of the polypeptides, with no need for evolutionary "blueprints." It was traditionally expected that such models – if valid – would fully explain the protein folding process. The Boltzmann paradigm is based on a thermodynamic hypothesis that views the folding process as contingent upon the polypeptide chain's tendency to spontaneously seek out a global free-energy minimum.[1]

The *ab initio* protein structure prediction problem essentially boils down to proposing an *in silico* folding algorithm for a known sequence of amino acids. This algorithm should reflect a balance between the degrees of freedom and the constraints imposed upon adjacent amino acids in polypeptide chains. The available degrees of freedom determine the nature of the connection between each pair of



Published by Woodhead Publishing Limited, 2012

peptides (for instance, GLY residues tend to preserve many degrees of freedom upon binding, while PRO residues form relatively rigid bonds, with few degrees of freedom).

The "all-atom" model (as it is usually called) involves an accurate representation of each atom in the protein molecule. Clearly, the number of relevant atoms can become quite large, from several hundred in small chains up to tens of thousands in larger proteins. Since the presence of the aqueous solvent must also be taken into account, the total number of atoms involved in "all-atom" simulations may easily exceed 100 000. Another important consideration is the complexity of the functions that model the pairwise interactions between atoms. Such functions tend to generate conformational hyperspaces with large numbers of dimensions and a plethora of local minima. Given these problems, numerical models oriented towards protein structure prediction focus on:

- simplifying the structure itself (by reducing the number of relevant elements and the number of degrees of freedom);
- simplifying the ways of traversing conformational spaces, by lowering the number of local minima.

The structural simplification often comes in the form of the so-called "united atoms" model, where certain clusters of atoms are replaced with a single element with comparable physiochemical properties. This is applied, for example, to $CH_3$ and $CH_2$ moieties. Additionally, the concept of "effective atoms" is frequently invoked when side chains (or even peptide bonds) are discussed. The effective atom is usually placed at the geometric center of the group of atoms it represents, and its properties mimic those of the original amino acid. Other possible ways to streamline geometric



Published by Woodhead Publishing Limited, 2012

representations of proteins will be presented later in this chapter.

In order to predict plausible protein structures, the conformational subspace must be traversed in search of states that correspond to internal free-energy minima. From a computational viewpoint, the problem involves optimizing a function of many variables (each representing a specific interaction that affects the protein). Upon locating a minimum, the algorithm should be able to determine whether it is in fact the global minimum of the entire conformational space, or merely one of its local minima. Answering this question with absolute certainty calls for global optimization methods that remain beyond the reach of modern computational science. Nevertheless, a scan of the conformational space (i.e., the function that ascribes energy values to specific conformations) remains a necessary step of any simulation algorithm. A frequently adopted means of simplifying this procedure is to introduce a well-defined quantum (or step size) for all conformational changes, and to assume that atoms (actual or effective) can only be found at the nodes of a regular three-dimensional grid. Making such assumptions greatly reduces the computational complexity of the problem, but the optimization comes at a cost. Replacing amino acid side chains with effective atoms and discretizing the conformational space requires one to define a pseudo-energy function that models interactions between virtual (as opposed to physical) objects. Thus, each virtual object must be parameterized by a somewhat arbitrary set of coefficients in order to distinguish between its allowed and disallowed energy states.

Scanning the conformational space for an effective-atom representation produces a number of candidate structures, which, when translated back into the all-atom model, facilitate analysis of their actual physiochemical properties.



Published by Woodhead Publishing Limited, 2012

In other words, simulating the folding process requires a direct correspondence between any simplified representation of the polypeptide chain and its actual physical counterpart.

Yet another potential optimization method is to replace the conformation/energy function with a so-called "force field deformation," where the local minima of the initial function are progressively smoothed out (in a stepwise fashion) until only one minimum remains. Upon locating this minimum, the deformation process can be reversed, leading to a conformation that reflects the global minimum of the initial function.

This chapter presents four optimization models, based on various algorithms (geometric simplification, free-energy function simplification, conformational space simplification, and field deformation).

## 9.2 Simplifying the geometric model and the field function

The model described below was proposed by one of the winners of the 1998 CASP3 challenge; it relies on simplifying the geometric structure of each amino acid in the polypeptide chain and proposes a simple function to model steric interactions. In addition, a special coefficient is introduced to normalize the field function as a whole. The model, called UNRES, is based on the earlier ECEPP/3 model; it has proven highly successful and is still being developed. Figure 9.1 shows how UNRES simplifies the structure of the polypeptide chain and its side chains (represented by effective atoms).

As shown in Figure 9.1, each side chain (labeled SC) is represented by an effective atom, located at its geometric center. The chain has two degrees of freedom: $\alpha$ (the deviation from the peptide bond surface) and $\beta$ (the rotation about the



Published by Woodhead Publishing Limited, 2012

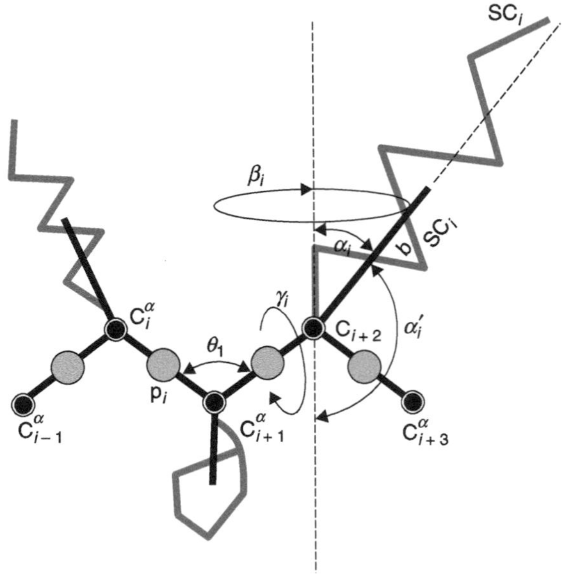


**Figure 9.1** **Schematic depiction of the structure of peptide bonds and their corresponding degrees of freedom according to UNRES**

side chain axis). In UNRES, each peptide bond is represented by an effective atom $p_i$ having degrees of freedom $\theta_i$ and $\gamma_i$, which correspond to rotations of the dihedral angles $\Phi$ and $\Psi$. The structural simplification of the protein molecule must be paired with an appropriate pseudo-energy function. In the model presented here, this function assumes the following form:

$$U = w_{\mathrm{SC}} \sum_{i<j} U_{\mathrm{SC}i\mathrm{SC}j} + w_{\mathrm{SCp}} \sum_{i \neq j} U_{\mathrm{SC}ipj} + w_{\mathrm{pp}}^{\mathrm{vdW}} \sum_{i<j-1} U_{pipj}^{\mathrm{vdW}}$$
$$+ w_{\mathrm{pp}}^{\mathrm{el}} \sum_{i<j-1} U_{pipj}^{\mathrm{el}} + w_{\mathrm{tor}} \sum_{i} U_{\mathrm{tor}}(\gamma_i) + w_{\mathrm{tord}} \sum_{i} U_{\mathrm{tord}}(\gamma_i, \gamma_{i+1})$$
$$+ w_{\mathrm{b}} \sum_{i} U_{\mathrm{b}}(\Theta_i) + w_{\mathrm{rot}} \sum_{i} U_{\mathrm{rot}}(\alpha_{\mathrm{SC}i}, \beta_{\mathrm{SC}i}, \Theta_i)$$
$$+ \sum_{m=3}^{4} w_{\mathrm{corr}}^{(m)} U_{\mathrm{corr}}^{(m)} + w_{\mathrm{bond}} \sum_{i=1}^{\mathrm{nbond}} U_{\mathrm{bond}}(d_i) + w_{\mathrm{SS}} \sum_{i} U_{\mathrm{SS}i} \qquad [9.1]$$



Published by Woodhead Publishing Limited, 2012

The UNRES approach relies on dividing interactions into those occurring between pairs of side chains (SC, including hydrophobic interactions), those occurring between side chains and peptides (SCp), and those which exist between pairs of peptides, in the context of both the electrostatic and the van der Waals forces. Changes in free energy are calculated for each value of $\theta$, taking account also of the contribution of disulfide bonds. An important element of the formula is a set of weights (labeled $w$). Their values (with a variable relation between them) allow the relative importance of each type of interaction to be studied.[1–3]

At the core of this pseudo-energy function lies the "corr" component, applicable to any structure in which the polypeptide fragment assumes a so-called "correlated" form. This occurs when the conformation is dependent not on a single amino acid but rather on a group of amino acids, which, when suitably arranged, produce a well-defined structural motif (this situation is also called "cooperative order"). Examples of "correlated" elements include, among others, $\beta$-twists and $\beta$-sheets. It should be noted that modeling such motifs *in silico* is notoriously unreliable (particularly in the case of $\beta$-sheets), as it requires the chain to assume a conformation that is inherently unstable and is not favored energetically when no hydrogen bonds can be formed with a partner polypeptide fragment. The role of $U_{corr}$ is therefore to "preserve" the emerging correlated structure during successive simulation steps. If one of the motifs associated with a "correlated" form begins to materialize, $U_{corr}$ can be adjusted to counteract any forces that could destroy it later on, when other conformational choices are possible for a given polypeptide.

As already suggested, introducing a pseudo-energy function in any simplified model calls for parameterization.



Published by Woodhead Publishing Limited, 2012

The parameter values do not reflect any physiochemical property (and thus cannot be based on experimental studies), but they can usually be approximated given sufficient expert knowledge of the phenomenon to be simulated and a means for quick iterative validation of the results. Applying a correctly parameterized function to the simplified conformational space produces prospective low-energy conformations that can be treated as an input to more accurate all-atom algorithms. These final steps are needed to model the free-energy distribution on the basis of actual physiochemical properties.

The UNRES model is being progressively extended by introducing specific "corr" potentials for various "correlated" forms.[4] A coarse-grained UNRES force field has been applied in molecular dynamics simulations.[5]

## 9.3 Lattice model

Another interesting approach to optimizing conformational-space traversal is known as the lattice model.[6,7] The authors of this model were attempting to simplify the representation of the polypeptide chain and the space in which the chain is being folded. Specifically, only $C_\alpha$ atoms are considered in the model, and it is assumed that a 105° angle exists between each pair of adjacent elements. This produces a tetrahedral lattice where $C_\alpha$ atoms are always found at the nodes. The conformations of side chains (represented by effective atoms) are determined by the rigid geometry of the core structure. The corresponding pseudo-energy function acknowledges the existence of hydrophobic interactions ($\varepsilon_h$), in addition to a so-called cooperativity criterion ($\varepsilon_c$), which leads to the emergence of a secondary structure.

In order to scan the conformational space quickly, the authors implemented an asymmetric Metropolis-based algorithm. This provides a somewhat crude but rapid assessment of the behavior and minima of the pseudo-energy function. Applying the lattice model to folding predictions involves several steps. First, a set of random conformations is generated for a known sequence of peptides. For each of these, a thermal Monte Carlo simulation is performed to obtain dynamic structures with the greatest possible variability. The molecular dynamics simulation is followed by simulated annealing. The results are clustered according to the values of their root mean square deviation (RMS-D) and the most densely populated clusters are translated into an all-atom model. This enables further optimization of the internal free energy, taking the complete structure of the polypeptide under consideration into account. The final structure is assumed to represent the expected result of the folding process for a specific initial sequence of amino acids. Figure 9.2 depicts a schematic view of the lattice algorithm (reduced to two dimensions for the sake of clarity).

Much like UNRES, the lattice model tries to simplify the chain structure by introducing virtual peptide bonds between pairs of $C_\alpha$ atoms and discretizing the search space. As already remarked, this approach only works if a suitable pseudo-energy function is supplied; however, establishing appropriate coefficients for this type of function remains a significant challenge. The advantage is a very fast search for an initial structure; the results can be quickly validated and the function adjusted to accurately mimic real-world processes. The lattice model is particularly well suited to generating the so-called contact matrix, which lists residues that remain in contact with each other in a fully folded protein molecule.



Published by Woodhead Publishing Limited, 2012

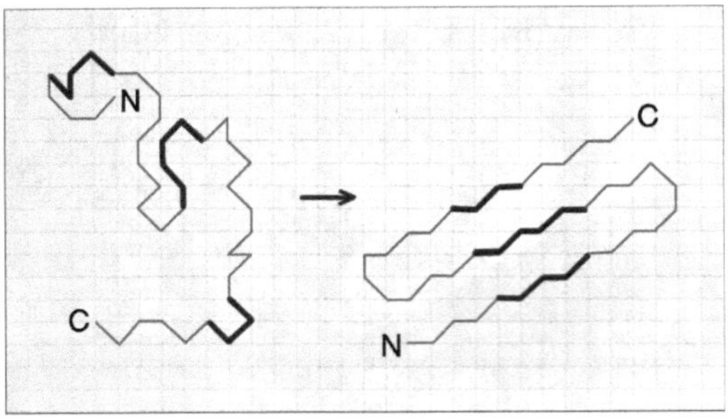


**Figure 9.2** Random starting structure (left) and the corresponding result generated using the lattice model (right). Only the locations of $C_\alpha$ atoms are depicted. Bold lines indicate hydrophobic residues (clearly aggregated in the right-hand image). The figure presents a two-dimensional view of the folding process

The lattice model has enjoyed considerable success in most of the CASP challenges, from CASP3 (1998),[8] through CASP4 (2000),[9] CASP5 (2002),[10] CASP6 (2004),[11] and CASP7 (2006),[12] to CASP8 (2008).[13]

## 9.4 ROSETTA

No review of protein structure prediction models would be complete without mentioning the ROSETTA package, one of the more successful participants in the CASP challenges.[14]

ROSETTA contains a custom database of three- and nine-residue segments. A set of structural motifs has been defined for each segment, and the simulation relies on assembling these motifs into a coherent 3D structure. Approximately 100 000 attempts are required for a typical target. The minimization of intramolecular interactions is based on a



Published by Woodhead Publishing Limited, 2012

special scoring function that takes into account hydrophobicity, disulfide bonds, sequence-independent terms, $\alpha$-helices, $\beta$-twists, and packed assemblies of $\beta$-sheets.[15]

ROSETTA assumes that local interactions can be modeled by mapping segments of the chain to distinct sets of structural motifs and that nonlocal interactions select the lowest-free-energy tertiary structures from the conformations compatible with the local mappings. Nonlocal interactions are modeled reasonably well at the level required for folding, without detailed representations of side chains.

ROSETTA-abinitio is an *ab initio* protein structure prediction software package based on the assumption that local interactions guide the conformation of short segments while global interactions determine the three-dimensional structure; this is also compatible with local biases.[16] To derive the local sequence/structure relationships for a given amino acid sequence (called the query sequence), ROSETTA-abinitio extracts from the PDB the distribution of conformations adopted by the query sequence segment in actual proteins.[16] Once these conformations are known, each query sequence may be included in a number of higher-order combinations using a Monte Carlo procedure. The resulting structures are subjected to energy minimization in a semiempirical force field[16] that takes account of hydrophobic and electrostatic interactions, core hydrogen bonds, and excluded volumes. Finally, the structures compatible with the local biases and nonlocal interactions are ranked according to their total energy calculated by the minimization procedure.

As already mentioned, ROSETTA-abinitio has been quite successful in successive CASP challenges, including CASP2 (1996),[17] CASP3 (1998),[18] CASP4 (2000),[19] CASP5 (2002),[20] CASP6 (2004),[21–24] CASP7 (2006),[25,26] and CASP8 (2008).[27,28] More information about this toolkit, along



Published by Woodhead Publishing Limited, 2012

with guidelines on how to perform custom simulations, can be found in the work of the Baker Laboratory.[29]

The EUChinaGrid project used ROSETTA to model a set of proteins called "never born proteins." Approximately 100 00 of these proteins (60 random amino acids each) were generated, assuming equal probability of occurrence of each amino acid. The set was expected to contain many pharmacologically active proteins. The protein structures were folded using both ROSETTA and the "fuzzy oil drop" model, yielding pairwise differences (RMS-D) of approximately 6.7–7.7 Å for structures with the greatest similarity and 25–32 Å for the most dissimilar structures. Comparative analysis of the two groups indicated the presence of a well-defined hydrophobic core in the structures produced by the "fuzzy oil drop" model (which was in line with expectations). This phenomenon was not consistently reproduced by ROSETTA: while the folded proteins did occasionally include highly hydrophobic zones, these zones were not always located near the center of the globular molecule.

Comparing the proteins with high RMS-D values leads to interesting conclusions: in some cases, ROSETTA produced globular structures while the "fuzzy oil drop" model came up with elongated fibrils, similar to almost fully extended polypeptide chains with no tertiary characteristics. It seems that the reason why ROSETTA always produces globular forms (even for nonexistent proteins, with distorted probabilities of occurrence of each amino acid) is its reliance on three- and nine-peptide segments, corresponding to structural motifs stored in ROSETTA's internal database. Assembling a molecule from such "preengineered" units always leads to a packed structure, even when the sequence itself is not found in nature (owing to the uneven populations of the various amino acids in natural proteins). In contrast,



Published by Woodhead Publishing Limited, 2012

the "fuzzy oil drop" model represents a holistic approach and performs optimization on a whole-molecule level – it can accurately reflect situations where an overabundance of certain peptides prevents the chain from assuming a globular form.

Detailed descriptions of the structures mentioned in this section can be found in Refs 30 and 31.

## 9.5 In search of a global minimum – force field deformation

Finding the global minimum of the free-energy function (which describes the intramolecular interactions) is a crucial step in any thermodynamic approach to protein folding. Such models typically posit that proteins "spontaneously" seek global energy minima. Since a fully reliable algorithm for global optimization of complex functions remains beyond the reach of computational science, many algorithms try instead to simplify the search space before actually commencing the search.

One such algorithm is known as "force field deformation." This relies on the properties of the second derivative, which can be added to the original function without altering its inflection points. This operation results in "smoothing out" of the function, including the force field function in particular. Shallow minima and gentle maxima can be eliminated altogether, leaving only "significant" peaks and troughs. Iterating this process leads gradually to a function with only one minimum, assumed to correspond to the global minimum of the original function. In formal terms, the procedure can be described as follows.

A small fraction of the second derivative is added to the original function $F^{[0]}(x)$:



Published by Woodhead Publishing Limited, 2012

$$F^{[1]}(x) = F(x)+\beta F''(x) \quad [9.2]$$

The process is performed iteratively:

$$F^{[2]}(x) = \left(1+\beta\frac{d^2}{dx^2}\right)F^{[1]}(x) \quad [9.3]$$

The process converges to the following deformation:

$$F^{[N]}(x) = \left(1+\beta\frac{d^2}{dx^2}\right)^N F(x) \quad [9.4]$$

All operations must be performed for a suitably small $\beta$, which can be defined as

$$\beta = \frac{t}{N} \quad [9.5]$$

Assuming that $t$ is small in relation to $N$, we obtain

$$F(x,t) = \lim_{N\to\infty}\left(1+\frac{t}{N}\frac{d^2}{dx^2}\right)^N F(x) = (e^{t\,d^2/dx^2})F(x) \quad [9.6]$$

If $F^{[N]}$ converges, $F(x, t)$ approximately solves the diffusion formula:

$$\frac{\partial^2 F(x,t)}{\partial x^2} = \frac{\partial F(x,t)}{\partial t} \quad [9.7]$$

where $t$ is a measure of the elapsed time, and $F(x, 0) = F(x)$. Figure 9.3 illustrates how the algorithm can be iterated to produce a function with a single global minimum.

It should be noted that the successive deformations displace the global minimum. Once the global minimum of the deformed function is located, an iterative process is needed to retrace the locus of the original function's minimum. Owing to the high complexity of the functions that represent interactions in real proteins, this procedure may prove cumbersome and lead to incorrect results. The problem of the displacement of the position of the global minimum can



Published by Woodhead Publishing Limited, 2012

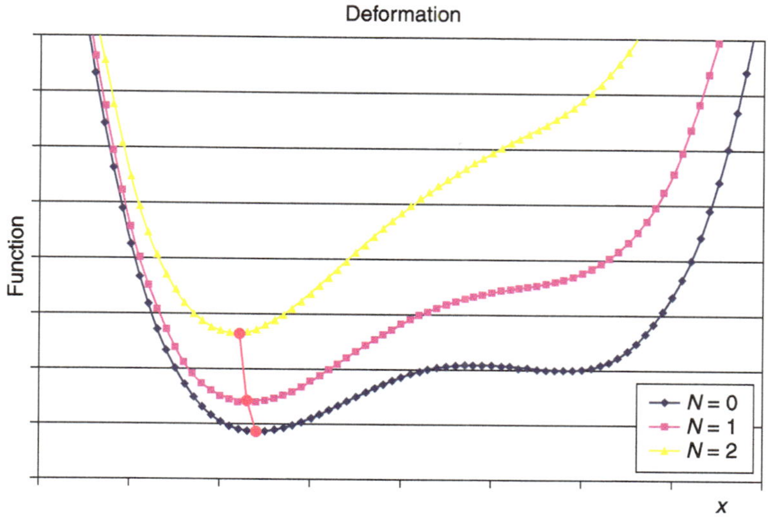


**Figure 9.3** Graphical depiction of deformation of the energy field by stepwise addition of "small portions" of the second derivative. Local minima are eliminated altogether, while the global minimum becomes displaced (as indicated by the red dots). In this example, the original function is a fourth-order polynomial and $\beta = 0.05$

be solved by applying a reversible procedure that recreates the deformation procedure. Nevertheless, an important conclusion can be drawn from a mathematical analysis (where Eq. 9.6 is reduced to a Taylor series) and from physical studies of the diffusion process: the model of $F(x, t)$ presented here eliminates narrow minima quickly in favor of broad troughs.

To better illustrate this phenomenon, let us consider a metal rod subjected to differential cooling as described in Figure 9.4(a), where a significant portion of the left-hand side is cooled to a moderate degree, while the central section contains a "spike-like" area where the cooling effect is greatly pronounced. Over time, the slight cooling of the end



Published by Woodhead Publishing Limited, 2012

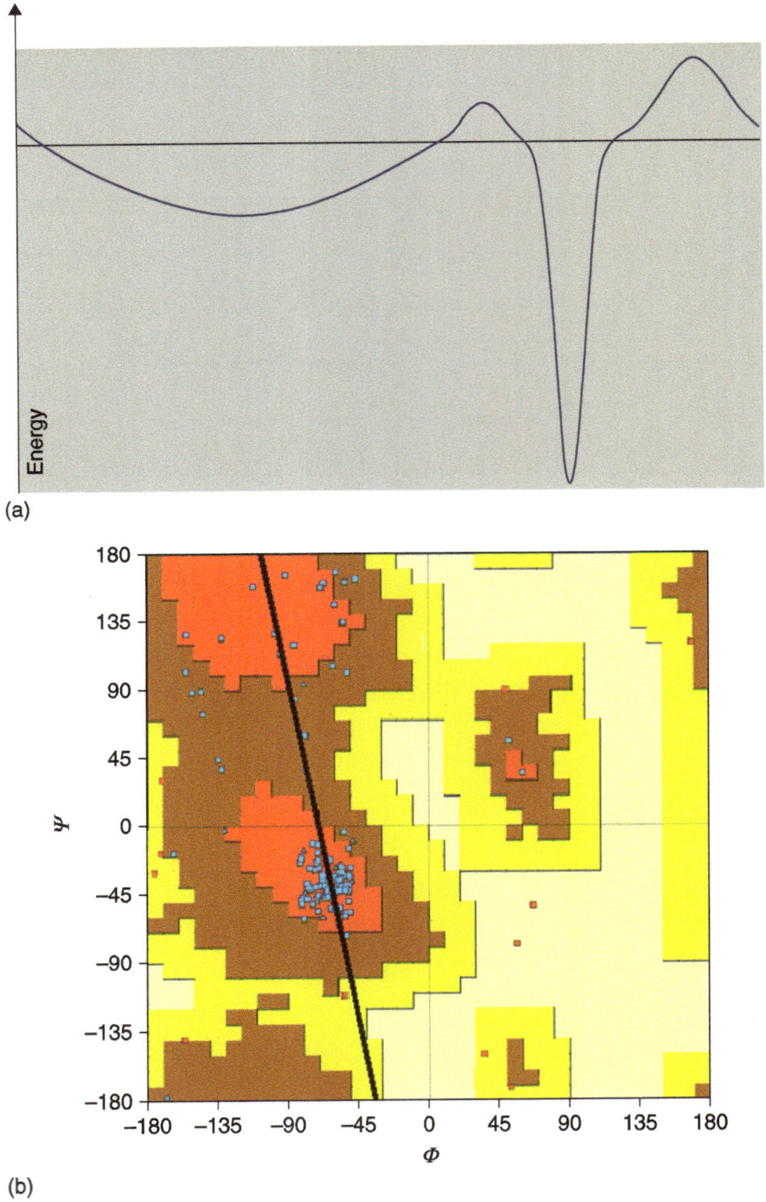


**Figure 9.4** Caption opposite



Published by Woodhead Publishing Limited, 2012

section will persist, whereas the central spike will be entirely eliminated. We should now mention that the initial distribution of thermal energy in the rod was not selected arbitrarily – rather, it reflects the distribution of internal energy in a particular conformational space for protein structures (see Figure 9.4(b)). It corresponds roughly to the spectrum of energy levels of structures from the $\beta$-structural area and the a-helix. The results of this thermal-diffusion simulation, where a wide, not necessarily low-temperature region is stable, in contrast to a short fragment cooled to a very low temperature, which disappears rapidly, can be applied to the relation between the $\alpha$ and $\beta$ conformations. The competition between these two conformations follows the relation between the deep, narrow minimum and the shallow, wide minimum for the two configurations of the polypeptide chain. According to the diffusion model, the entropic processes that govern folding should favor broad troughs ($\beta$ motifs) over deep but narrow spikes ($\alpha$-helices).

The model presented here shows that polypeptide chain folding, when interpreted on the basis of diffusion

**Figure 9.4** **Distribution of temperature along a differentially cooled rod. According to the laws of diffusion, reheating will increase the temperature along the entire length of the rod, but narrow minima will disappear much faster than broad, shallow troughs. The initial distribution reflects the distribution of energy in a $\beta$-to-$\alpha$ connector, as depicted in the diagram at the bottom. The rapid disappearance of narrow ($\alpha$-helical) motifs suggests that the relation between $\alpha$-helices and the comparatively longer-life $\beta$-twists needs to be artificially biased. The bottom map is taken from PDBsum**



Published by Woodhead Publishing Limited, 2012

phenomena, must represent a middle ground between energetic and entropic states. The free-energy function must be biased to favor helical motifs (deep global minima) over shallow but persistent troughs that correspond to $\beta$-twists. As both of these forms occur in actual proteins, there is a clear need for a compromise solution that reconciles the properties of $\alpha$ and $\beta$ structural motifs (i.e., a balance between energetic and entropic minima). A detailed description of the model presented here can be found in Refs 32 to 38.

## 9.6 References

### *9.6.1 UNRES model*


1. Bystroff C, Shao Y. Modeling protein folding pathways. In: Practical Bioinformatics Ed. Janusz M. Bujincki–15 Nucleic Acids and Molecular Biology vol. 15. Series Ed. Gross HJ. Springer–2004, pp 97–122.
2. Liwo A., Arłukowicz P., Czaplewski C., Ołdziej S., Pillardy J., Scheraga H.A. (2002) A method for optimizing potential-energy functions by a hierarchical design of the potential-energy landscape: application to the UNRES force field. *Proc. Natl. Acad. Sci. USA* 99(4), 1937–42.
3. Liwo A., Pincus M.R., Wawak R.J., Rackovsky S., Scheraga H.A. (1993) Prediction of protein conformation on the basis of a search for compact structures: test on avian pancreatic polypeptide. *Protein Sci.* 2(10), 1715–31.
4. Ołdziej S., Czaplewski C., Liwo A., Chinchio M., Nanias M., et al. (2005) Physics-based protein-structure prediction using a hierarchical protocol based on the UNRES force field: assessment in two blind tests. *Proc. Natl. Acad. Sci. USA* 102(21), 7547–52.




Published by Woodhead Publishing Limited, 2012

5. Maisuradze G.G., Senet P., Czaplewski C., Liwo A., Scheraga H.A. (2010) Investigation of protein folding by coarse-grained molecular dynamics with the UNRES force field. *J. Phys. Chem. A* 114(13), 4471–85.
6. Liwo A., Ołdziej S., Czaplewski C., Kleinerman D.S., Blood P., Scheraga H.A. (2010) Implementation of molecular dynamics and its extensions with the coarse-grained UNRES force field on massively parallel systems; towards millisecond-scale simulations of protein structure, dynamics, and thermodynamics. *J. Chem. Theory Comput.* 6(3), 890–909.


## 9.6.2 Lattice model


7. Kolinski A., Skolnick J., Yaris R. (1987) Monte Carlo studies on equilibrium globular protein folding. I. Homopolymeric lattice models of beta-barrel proteins. *Biopolymers* 26(6), 937–62.
8. Skolnick J., Kolinski A., Yaris R. (1988) Monte Carlo simulations of the folding of 3-barrel globular proteins. *Proc. Natl. Acad. Sci. USA* 85, 5057–61.
9. Ortiz A.R., Kolinski A., Rotkiewicz P., Ilkowski B., Skolnick J. (1999) Ab initio protein structure prediction using restraints derived from evolutionary information. *Proteins* S3, 177–85.
10. Skolnick J., Kolinski A., Kihara D., Betancour M., Rotkiewicz P., Boniecki M. (2001) Ab initio protein structure prediction via a combination of threading, lattice folding, clustering and structure refinement. *Proteins* S5, 149–56.
11. Skolnick J., Zhang Y., Arakaki A.K., Kolinski A., Boniecki M., et al. (2003) TOUCHSTONE: a unified approach to protein structure prediction. *Proteins* S6, 469–79.




Published by Woodhead Publishing Limited, 2012

12. Zhang Y., Arakaki A.K., Skolnick J. (2005) TASSER: an automated method for the prediction of protein tertiary structures in CASP6. *Proteins* 61(S7), 91–8.
13. Zhou H., Pandit S.B., Lee S.Y., Borreguero J., Chen H., et al. (2007) Analysis of TASSER-based CASP7 protein structure prediction results. *Proteins* 69(S8), 90–7.
14. Zhou H., Pandit S.B., Skolnick J. (2009) Performance of the Pro-sp3-TASSER server in CASP8. *Proteins* 77(S9), 123–7.

### 9.6.3 *ROSETTA*

15. Protein Structure Prediction Center. Available from: *http://predictioncenter.org/* [accessed 7 May 2012].
16. Simons K.T., Ruczinski I., Kooperberg C., Fox B.A., Bystroff C., Baker D. (1999) Improved recognition of native-like protein structures using a combination of sequence-independent features of proteins. *Proteins* 34, 82–95.
17. Rohl C.A., Strauss C.E., Misura K.M.S., Baker D. (2004) Protein structure prediction using Rosetta. *Methods Enzymol.* 383, 66–93.
18. Bystroff C., Baker D. (1997) Blind predictions of local protein structure in CASP2 targets using the I-sites library. *Proteins* Suppl. 1, 167–71.
19. Simons K.T., Bonneau R., Ruczinski I., Baker D. (1999) Ab initio protein structure prediction of CASP III targets using ROSETTA. *Proteins* Suppl. 3, 171–6.
20. Bonneau R., Tsai J., Ruczinski I., Chivian D., Rohl C., et al. (2001) Rosetta in CASP4: progress in ab initio protein structure prediction. *Proteins* Suppl. 5, 119–26.
21. Chivian D., Kim D.E., Malmström L., Bradley P., Robertson T., et al. (2003). Automated prediction of

CASP-5 structures using the Robetta server. *Proteins* 53 (Suppl. 6), 524–33.

22. Chivian D., Kim D.E., Malmström L., Schonbrun J., Rohl C.A., Baker D. (2005) Prediction of CASP6 structures using automated Robetta protocols. *Proteins* 61 (Suppl. 7), 157–66.
23. Kim D.E., Chivian D., Malmström L., Baker D. (2005) Automated prediction of domain boundaries in CASP6 targets using Ginzu and RosettaDOM. *Proteins* 61 (Suppl. 7), 193–200.
24. Graña O., Baker D., MacCallum R.M., Meiler J., Punta M., et al. (2005) CASP6 assessment of contact prediction. *Proteins* 61 (Suppl. 7), 214–24.
25. Bradley P., Malmström L., Qian B., Schonbrun J., Chivian D., et al. (2005) Free modeling with Rosetta in CASP6. *Proteins* 61 (Suppl. 7), 128–34.
26. Das R., Qian B., Raman S., Vernon R., Thompson J., et al. (2007) Structure prediction for CASP7 targets using extensive all-atom refinement with Rosetta@home. *Proteins* 69 (Suppl. 8), 118–28.
27. Tress M., Cheng J., Baldi P., Joo K., Lee J., et al. (2007) Assessment of predictions submitted for the CASP7 domain prediction category. *Proteins* 69 (Suppl. 8), 137–51.
28. Krieger E., Joo K., Lee J., Lee J., Raman S., et al. (2009) Improving physical realism, stereochemistry, and side-chain accuracy in homology modeling: four approaches that performed well in CASP8. *Proteins* 77 (Suppl. 9), 114–22.
29. Raman S., Vernon R., Thompson J., Tyka M., Sadreyev R., et al. (2009) Structure prediction for CASP8 with all-atom refinement using Rosetta. Proteins 77 (Suppl. 9), 89–99.
30. Baker Laboratory. Available from: *http://depts.washington.edu/bakerpg/* [accessed 7 May 2012].



Published by Woodhead Publishing Limited, 2012

31. Prymula K., Piwowar M., Kochanczyk M., Flis L., Malawski M., et al. (2009) In silico structural study of random amino acid sequence proteins not present in nature. *Chem. Biodivers.* 6(12), 2311–36.
32. Minervini G., Evangelista G., Villanova L., Slanzi D., De Lucrezia D., et al. (2009) Massive non-natural proteins structure prediction using grid technologies. *BMC Bioinform.* 10 (Suppl. 6), S22.

### 9.6.4 Force field deformation

33. Piela L., Kostrowicki J., Scheraga H.A. (1989) The multiple-minima problem in the conformational analysis of molecules. Deformation of the potential energy hypersurface by the diffusion equation method. *J. Phys. Chem.* 93, 3339–46.
34. Kostrowicki J., Piela L., Cherayil B.J., Scheraga H.A. (1991) Performance of the diffusion equation method in searches for optimum structures of clusters of Lennard-Jones atoms. *J. Phys. Chem.* 95, 4113–19.
35. Kostrowicki J., Scheraga H.A. (1992) Application of the diffusion equation method for global optimization to oligopeptides. *J. Phys. Chem.* 96, 7442–9.
36. Kostrowicki J., Scheraga H.A. (1995) Simple global minimization algorithm for one-variable rational functions. *J. Global Optim.* 6, 293–311.
37. Kostrowicki J., Scheraga H.A. (1995) New method for calculation of the exact end-to-end distance distributions for the freely-rotating chain. *Comput. Polym. Sci.* 5, 47–55.
38. Scheraga H.A., Hao M.H., Kostrowicki J. (1995) Theoretical studies of protein folding. In Atassi M.Z. and Appella E. (eds), *Methods in Protein Structure Analysis*, pp. 457–64. New York: Plenum.



Published by Woodhead Publishing Limited, 2012

39. Kostrowicki J., Scheraga H.A. (1996) Some approaches to the multiple-minima problem in protein folding. In Pardalos P.M., Shalloway D., and Xue G. (eds), *Global Minimization of Nonconvex Energy Functions: Molecular Conformation and Protein Folding*, pp. 123–32, DIMACS Series in Discrete Mathematics and Theoretical Computer Science No. 23. Providence, RI: American Mathematical Society.

## 9.7 Suggested reading

The models presented here can be considered as the most effective and widespread approaches to protein structure prediction currently available. As their authors are unquestioned leaders in the field (the group includes, among others, A.H. Scheraga, J. Skolnick, D. Baker, M. Levitt, V. Daggett, and M. Karplus), citing their papers here is unnecessary. Any search engine (such as PubMed) will readily produce an extensive list of publications by each author's team, with detailed descriptions of their models, algorithms, and case studies.

It is strongly suggested that the CASP (Critical Assessment of protein Structure Prediction) webpage is visited. A comprehensive review analysis of the worldwide progress in protein structure prediction can be monitored there.

## 9.8 Selected servers oriented toward protein structure prediction available on the Internet

*I-TASSER* is the best service for protein structure prediction, according to the 2006–2010 CASP experiments (CASP7,



Published by Woodhead Publishing Limited, 2012

CASP8, and CASP9). The standalone I-TASSER package is freely available for download.

*RaptorX* excels at aligning hard targets, according to the 2010 CASP9 experiments. RaptorX generates significantly better alignments for the hardest 50 CASP9 template-based modeling targets than do other servers, including those using consensus and refinement methods. RaptorX is available as a service.

*MODELLER* is a popular software tool for producing homology models using a methodology derived from the processing of NMR spectroscopy data. SwissModel provides an automated web service for basic homology modeling.

*HHpred*, *bioinfo.pl*, and *Robetta* are widely used for protein structure prediction. HHsearch is a free software package for protein threading and remote homology detection.

*SPARKSx* is one of the top-performing tools in CASP, focused on remote fold recognition.

*PEP-FOLD* is a *de novo* approach aimed at predicting peptide structures from amino acid sequences, based on the HMM structural alphabet.

*Phyre* and *Phyre2* are among the top-performing tools in the CASP international blind trials of structure prediction in homology modeling and remote fold recognition. Both are designed with an emphasis on ease of use for nonexperts.

*RAPTOR* is a software package for protein threading that is based on integer programming.

*QUARK* is an online service suitable for *ab initio* protein structure modeling.



Published by Woodhead Publishing Limited, 2012

*Abalone* is a molecular dynamics program for folding simulations with explicit or implicit water models.

*TIP* is a knowledge base of STRUCTFAST models and precomputed similarity relationships between sequences, structures, and binding sites.

Distributed computing projects concerning protein structure prediction have also been implemented, such as Folding@home, Rosetta@home, the Human Proteome Folding Project, Predictor@home, and TANPAKU.

The FOLDIT program seeks to investigate the pattern recognition and puzzle-solving abilities inherent in the human mind in order to create more successful software for computer protein structure prediction.



Published by Woodhead Publishing Limited, 2012

# Conclusion

*L. Konieczny and I. Roterman-Konieczna*

**Abstract:** The multistage *in silico* protein folding model presented in this work enables folding processes to be simulated on the basis of amino acid sequences, taking several new factors into account, namely (1) the introduction of an external force field expressing the presence of a water solution; (2) the existence of different optimization procedures, i.e., pairwise optimization of internal interactions (electrostatic, van der Waals, and torsion-potential forces, all of which follow a random distribution) and molecular-level optimization of external interactions (which affect the molecule as a whole and result in the creation of a hydrophobic core); and (3) the presence of ligands in close proximity to the folding chain, affecting the structure of the resulting protein and ensuring its high specificity. Acknowledging these phenomena (which are not accurately represented in existing models) leads to a reinterpretation of the entire protein folding process.

**Key words:** conformational space for protein folding, multistage process, multicriteria optimization, internal force field, external force field, ligand binding



Published by Woodhead Publishing Limited, 2012

Molecular dynamics simulations (frequently used to model changes in biomolecular structures), while firmly grounded in physics, do not fully reflect the properties of many biological processes. High-temperature simulations allow the modeling of structural changes that occur in response to an influx of energy, but do not explain the biological consequences of such changes. Exposing a protein molecule to an energy source is a directionless process – unlike the generation of enzymatic active sites or ligand-binding pockets, which requires an external source of information. Thus, in general, the analysis of postdynamics structures only provides insight into the distribution of static areas (where residues possess few degrees of freedom as a result of their structural arrangement) and dynamic areas (where local conditions promote greater conformational variability). Such data may sometimes contain hints about the biological activity of a protein, but this is not always the case.

Structural changes within a protein molecule must therefore be treated as a reaction to outside stimuli. External factors (ligands, membranes, ions, etc.) may induce local changes in the protein's structure, resulting in steric and entropic adaptation to the changing environment. According to our proposed model, a polypeptide chain, when immersed in pure water (although it should be noted that the presence of ions is always required to ensure an appropriate pH), undergoes folding that can be accurately modeled by the "fuzzy oil drop" paradigm. The degree to which actual (observed) protein structures deviate from this idealized state depends on the influence of the external environment. Analysis of postdynamics structures (formed under high-temperature conditions) points to the areas that are most susceptible to being modified by external factors; however, the actual changes are not simply a reaction to an influx of thermal energy.



Published by Woodhead Publishing Limited, 2012

Information theory clearly indicates that the quantity of information carried by a sequence of amino acids is only sufficient to generate the early-stage (ES) folding intermediate and that additional information must be provided in order to correctly model the remainder of the folding process. We postulate that this information comes from the external environment, which plays a crucial role in determining the final shape of the protein. In any case, determining the ES structure (in whatever manner that may be feasible) remains crucial as a starting point for free-energy minimization algorithms. It is worth noting that our proposed model provides such a structure, along with a clear explanation of how it is obtained and what mechanisms are responsible for this.

In-depth analysis based on the concepts of information theory has led us to partition the Ramachandran map into areas that correspond to specific structural motifs, including two subclasses of $\beta$-sheets (E and F) and three subclasses of random coils (A, B, and D). In light of the proposed model, special attention should be devoted to classes F and D. Their proximity to $\beta$-sheets (E) and right-handed helices (C) on the one hand and their specific contribution to the final native structure (initiation and/or termination of $\beta$-sheets and helical fragments) on the other suggest that they play a critical role in determining the shape of the folded protein. Molecular dynamics simulations indicate that $\beta$-sheets form easily without the need for any primers (or correlated potentials), which in turn highlights the importance of structures belonging to zone F. Thus, if the early-stage intermediate allows us to pinpoint D and F structures (in relation to C and E motifs), it is highly probable that we shall correctly predict the entire secondary structure of the polypeptide chain.



Published by Woodhead Publishing Limited, 2012

The final form of the protein is dependent on external force fields, which induce the chain to fold in a manner consistent with its intended biological role. Environmental influences may manifest themselves through forces exerted by water (resulting in a 3D Gaussian-like distribution of hydrophobicity) or through factors other than the solvent itself, which interact with the chain and promote specific folding patterns.

The need to take external factors into account was already evident in CASP5,[1] where the structure of T0142 (PDB code 1NTF)[2] was found to be so highly dependent on the presence of a ligand that this fact was communicated to contest participants. In this specific case, it was felt that no contestant could obtain the large flat sheet (required to bind heme) using traditional methods. Applying the "fuzzy oil drop" model to 1NTF (in its complexed state) yields *O/T* and *O/R* values of 0.205 and 0.192, respectively, suggesting that its actual hydrophobic core does not mirror theoretical predictions. The elimination of residues responsible for binding ligands has a negligible influence on the *O/T* and *O/R* values; however, eliminating the NO-binding CYS residue at position 60 (together with its immediate surroundings) reduces *O/T* to 0.193 while increasing *O/R* to 0.239. It seems that the ligand itself is not responsible for the deformations in the protein's hydrophobic core – rather, the significant deformation comes from its "structurally encoded" biological properties, which call for a specific pattern of binding of NO molecules.

Another important conclusion is that energy optimization occurs under diverse conditions and that correctly modeling this process requires a multicriteria optimization approach.

The multicriteria optimization approach can be generalized as follows:

$$F(r_{ij}) = [F_{\mathrm{INT}}(r_{ij}), F_{\mathrm{EXT}}(r_{ij})]^T$$

where $F_{\mathrm{INT}}(r_{ij})$ is the internal force field, $F_{\mathrm{EXT}}(r_{ij})$ is the external force field, and $r_{ij}$ is the distance between the interacting elements (actual or effective atoms). The function $F(r_{ij})$ determines the aggregate energy state for a given object (the protein), with contributions from $F_{\mathrm{INT}}(r_{ij})$ (which expresses the internal interactions) and $F_{\mathrm{EXT}}(r_{ij})$ (which corresponds to deformations triggered by external factors). The optimization of $F_{\mathrm{INT}}(r_{ij})$ (called LI optimization in Chapter 7), which can be described as seeking a state of suitably low energy for nonbinding interactions, converges to a minimum of this constituent function. Calculations are performed using an all-atom model, as the interactions between each pair of atoms must be considered. This approach eliminates the need for a simplified model (the geometry and conformational space of the molecule are fully represented); thus there is no need to define a pseudo-energy function.

In contrast, the optimization of $F_{\mathrm{EXT}}(r_{ij})$ (called TM optimization in Chapter 7) relies on a different convergence criterion – namely, the minimization of the differences between the observed and expected distributions of hydrophobicity in the protein molecule. The actual distribution is influenced by water (represented by a 3D Gaussian function) and, optionally, by ligands, which distort the idealized structure of the emerging hydrophobic core. In these calculations, the protein's geometry is represented by effective atoms (the geometric centers of each side chain). The only parameterization required for $F_{\mathrm{EXT}}(r_{ij})$ is the proper choice of the hydrophobicity scale, which is available in many forms (theoretical and experimental). Any scale can be adapted to these calculations. The ultimate goal of multicriteria optimization is finding an optimal energy state that takes account of both types of interactions (internal and external).

To summarize, the following observations can be made at this point:

1. When the early-stage intermediate is modeled, the quantity of information carried by the amino acid sequence is only sufficient to select a structure from the limited conformational subspace. It seems that an ES model is sufficient for secondary-structure generation, whereas a model of the late-stage (LS) step is necessary for the construction of the tertiary structure.
2. Determining the final structure (in the complete conformational space) requires additional information, which is provided mostly by the water solvent. The presence of water affects the side chain conformations, producing a hydrophobic core whose shape reflects its environment. It is a well-known fact that the properties of various proteins may depend dramatically on the external conditions, up to the point of denaturation. In such situations, removing the denaturing factor often causes the protein to revert to its native state. In the context of our model, the presence of a denaturing factor manifests itself as a specific deformation of the hydrophobicity field, exerting a specific influence on the protein structure. It is important to underline that the most popular denaturant, urea (and also the precipitant $(NH_4)_2SO_4$), does not interact with proteins. Urea, at a suitably high concentration, engages water in its net organization. In consequence, the water is disordered to such an extent that the probability of generating a "fuzzy oil drop" is excluded. As a result, the water environment is not able to maintain a regular "fuzzy oil drop" system for the protein molecule. Stepwise removal of the



Published by Woodhead Publishing Limited, 2012

denaturing factor allows reversible generation of the "fuzzy oil drop" organization, ensuring the proper folding.

3. Additional information (required for determining the final structure) may be provided by external ligands whose presence at the folding stage results in the creation of a highly specific binding pocket.
4. The generation of a highly specific binding pocket can be interpreted as a specific deformation of the protein's hydrophobic core, enabling the protein to perform its intended biological function with high specificity.
5. Useful simulations of the protein folding process must involve a multicriteria optimization that reconciles nonbinding interactions (local, producing uniform conditions throughout the protein body) with forces acting on the whole molecule (e.g., by optimizing the differences between the actual and theorized structures of the protein's hydrophobic core). This is depicted schematically in Figure 10.1.
6. The presence of external factors and their participation in the protein folding process are treated as supporting the main goal of the protein folding process, of folding to obtain the correct expected biological function, not just the correct expected structure, as nature is interested in function and specific activity. The structure is only the material basis (or tool) that ensures that the main goal – specific activity – is achieved.
7. The presence of external factors may be limited to the presence of water, although it may be extended to include any other molecule within the interaction distance. The list of external factors increases when the complete "history" of the folding is taken into consideration, including interaction with a membrane, for example.



Published by Woodhead Publishing Limited, 2012

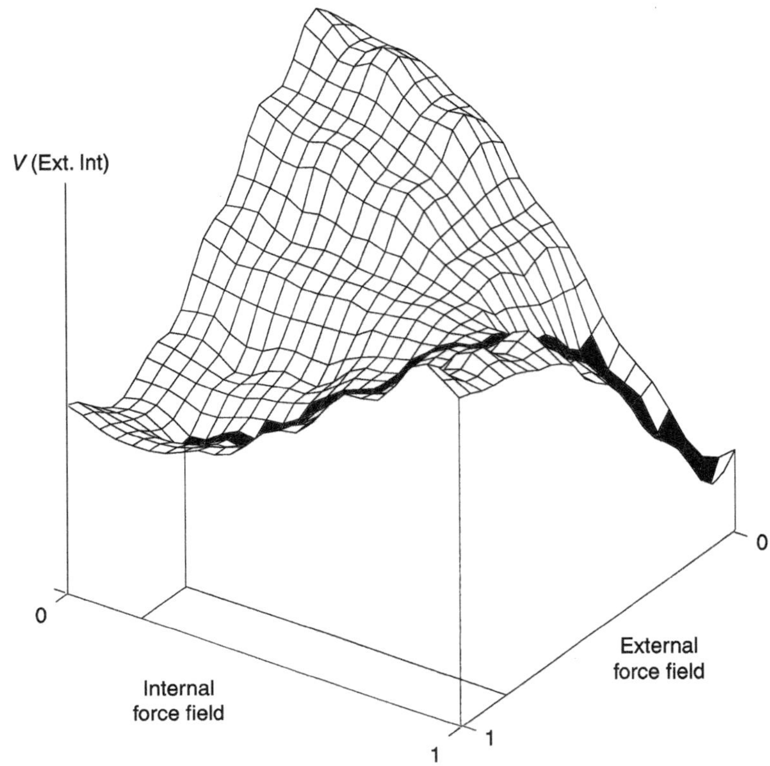


**Figure 10.1** Schematic depiction of the conformation space, taking account of the influence of external factors. Optimization of nonbinding interactions leads to a low-energy state (where the participation of the external force field is nil), resulting in an internally optimized molecule. However, as the influence of external factors increases,the internal energy state begins to deviate from its theoretical global optimum. As shown here, a consensus has been found between the two tendencies. Note that the graph presented here is meant as an example only and may not correspond to the actual distribution of energy in a protein molecule



Published by Woodhead Publishing Limited, 2012

8. The role of particular fragments of a polypeptide is especially important in the context of the folding "history" when some regions of the protein molecule play a functional role in transport of the molecule (through a membrane, for example) using the N-terminal fragment as the "leader."

To summarize the concepts presented in this book, we can refer once again to the folding schema given in the Introduction. Assuming that this process consists of several stages, the most challenging step involves the modeling of the LS structure (Figure 10.2). The LS model must take account of external factors (such as the ubiquitous presence of water, represented by a 3D Gaussian function), as well as interaction with ligands, other protein molecules, or potential substrates for enzymatically active proteins. Each of these factors is associated with a specific deformation of the external force field.

As hinted in the Introduction, the model is based on experimental evidence, particularly with regard to the division of the folding process into specific intermediates with well-known properties. The ES is dominated by the conformational preferences of individual amino acids, resulting in a backbone structure. It does not involve side chains (at least not explicitly) – instead, the properties of the various amino acids are reflected by relating them to fragments of the elliptical path that determines the limited

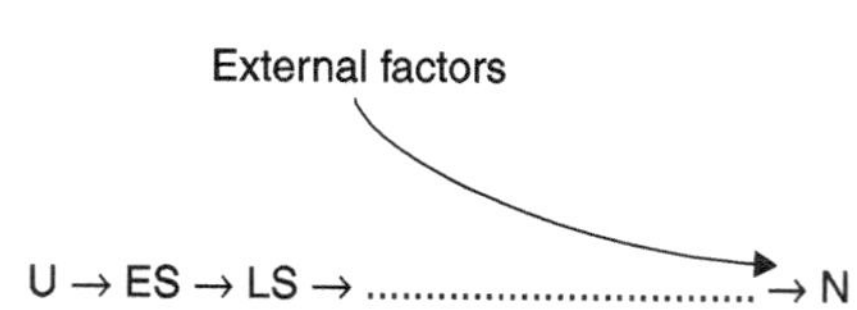


**Figure 10.2** The folding schema, taking account of external factors



Published by Woodhead Publishing Limited, 2012

conformational subspace, and expressing the conformational preferences of the amino acids using letter codes, depending on which fragments of the path they are most likely to belong to.

The LS folding depends on the external environment, described by a 3D Gaussian function. In certain cases this function alone is enough to accurately describe the protein structure, with a well-formed hydrophobic core at its center. This is evident, for example, in fast-folding (downhill) and antifreeze proteins. Such proteins typically exhibit a limited activity profile and do not require intricate, specific active sites to perform their biological function. The CASP challenge seems to acknowledge this phenomenon by classifying proteins into "easy" and "difficult" sets.[3] In "easy" proteins, the folding process either terminates at the ES or does not depend on external stimuli other than water. "Difficult" proteins are usually characterized by a high presence of $\beta$-type structural motifs or highly specific structural elements that facilitate their biological function.

Our observations are supported further by concepts from information theory. The information content of the polypeptide chain is only sufficient to determine the ES folding intermediate – further steps in the folding process (particularly those related to the generation of an active site) require additional information, which can only come from the external environment. We are currently investigating several implementations of the model presented here, differing with respect to the diversity and characteristics of the various polypeptide chains, as well as the dynamic properties exhibited by some proteins and their fragments. Our prospective involvement in the CASP initiative will enable us to determine which version of the model is most suitable for simulating actual proteins.



Published by Woodhead Publishing Limited, 2012

Experimentally based suggestions concerning the participation of ribosomes and their influence on polypeptide chain conformations have been presented recently.[4,5]

Returning to the ideas expressed in the Introduction, one can say that the "fuzzy oil drop" model makes it possible to search for an answer to the question "Why do proteins fold the way they do?" At least, it is possible to define the reasons for and origin of the failures that occur in protein folding *in silico* based on the "fuzzy oil drop" model.

## 10.1 Acknowledgements

The research presented here was carried out from 1995 to 2011 and was funded by a series of grants from the Jagiellonian University Medical College. The CYFRONET AGH Kraków Academic Computing Center provided computational support. We would very much like to express our thanks to Piotr Nowakowski of CYFRONET AGH for valuable editorial remarks. Many thanks also to Anna Zaremba-Śmietańska for technical support.

## 10.2 References

1. Venclovas Č. (2003) Comparative modeling in CASP5: progress is evident, but alignment errors remain a significant hindrance. *Proteins* 53(S6), 380–8.
2. Weichsel A., Maes E.M., Andersen J.F., Valenzuela J.G., Shokhireva T.Kh., et al. (2005) Heme-assisted S-nitrosation of a proximal thiolate in a nitric oxide transport protein. *Proc. Natl. Acad. Sci. USA* 102, 594–9.
3. Orengo C.A., Bray J.E., Hubbard T., LoConte L., Sillitoe I. (1999) Analysis and assessment of ab initio



Published by Woodhead Publishing Limited, 2012

three-dimensional prediction, secondary structure, and contacts prediction. *Proteins* 37 (Suppl. 3), 149–70.
4. Kaiser C.M., Goldman D.H., Chodera J.D., Tinoco I. Jr., Bustamante C. (2011) The ribosome modulates nascent protein folding. *Science* 334, 1723–7.
5. Perskie L.L., Street T.O., Rose G.D. (2008) Structures, basins, and energies: A deconstruction of the protein coil library. *Protein Science* 17, 1151–61.

## 10.3 Suggested reading

Clark P.L. (2004) Protein folding in the cell: reshaping the folding funnel. *Trends Biochem. Sci.* 29(10), 527–34.



Published by Woodhead Publishing Limited, 2012

# Index





Published by Woodhead Publishing Limited, 2012

Published by Woodhead Publishing Limited, 2012

Published by Woodhead Publishing Limited, 2012

Published by Woodhead Publishing Limited, 2012

Published by Woodhead Publishing Limited, 2012

Published by Woodhead Publishing Limited, 2012

Kolibri 360 GmbH
Ferdinand-Jühlke-Straße 7
99095 Erfurt, Deutschland
produktsicherheit@kolibri360.de